Veniceland Atlantis

Veniceland Atlantis

THE BLEAK FUTURE OF
THE WORLD'S FAVOURITE CITY

ROBERT L. FRANCE

LIBRI
PUBLISHING

A Web-Enhanced Book

Published in 2011 by Libri Publishing ■ Copyright © Robert L France ■ ISBN: Paperback 978 1 907471 13 1 ISBN: Hardback 978 1 907471 31 5 ■ All rights reserved. ■ A CIP catalogue record for this book is available from The British Library ■ Design by Helen Taylor ■ Printed in the UK by Ashford Colour Press ■ Libri Publishing, Brunel House, Volunteer Way, Faringdon, Oxfordshire SN7 7YR ■ Tel: +44 (0)845 873 3837 ■ www.libripublishing.co.uk

FOR RUI, WHO FIRST INTRODUCED ME, FOR THE
MEMORY OF MY MOTHER, MY CICERONE FOR
THE NEXT VISIT, AND FOR THE CITIZENS WHO
CONTINUE TO LIVE THERE.

"IF HUMAN BEINGS CAN BUILD A CITY LIKE THIS,
THEIR SOULS DESERVE TO BE SAVED."

Anthony Burgess *in* Hughes, H.M. (Ed.)
Venice: A Collection of the Poetry of Place (2006)

Contents

Preface

VENICE, as we all know, or perhaps more accurately as we all believe we know, is sinking. This is certainly the case, at least with regard to it drowning under more than a century of gushingly romanticised and effusively mythologising verbiage from its many besotted admirers and ardent worshipers. As M. Lovric expresses it in her recent anthology about Venice:

> "Far too many visiting poets have slathered the city with the effluvia of their sentimental raptures." (Lovric, 2003)

Indeed, thousands of books exist singing the manifest praises of the "Queen of the Adriatic". Entire forests are sacrificed each and every year to support the latest crop of these efforts. M. McCarthy's remark with respect to Venice, made a half century ago, that:

> "Nothing can be said here (including this statement) that has not been said before" (McCarthy, 1963)

is apt now more than ever – with one possible exception. Against this rising tide of ecstatic prose and pretty pictures, about a dozen books have been published that present a very different point of view.

Venice's environmental problems were first addressed for a popular audience in S. Fay and P. Knightly's *The Death of Venice* in 1976, and then more recently in J. Keahey's *Venice Against the Sea: A City Besieged* in 2002. Venice's equally significant social problems were touched upon in R. de Combray's *Venice, Frail Barrier: Portrait of a Disappearing City* in 1975, followed by P. Barbaro's *Venice Revealed: An Intimate Portrait* in 1998 and R. Debray's *Against Venice* in 1999.

Addressing Venice's environmental problems are other technical (and often expensive) books written by specialists and for scholars. These include P. Lasserre and A. Marzollo's *The Venice Lagoon Ecosystem: Impacts and Interactions Between Land and Sea*, 2000; C. Fletcher and J. Da Mosto's *The Science of Saving Venice*, 2004; and C. Fletcher and T. Spencer's *Flooding and Environmental Challenges for Venice and its Lagoon: State of Knowledge*, 2005. Environmental economic concerns are addressed in A. Alberni

et al.'s *Valuing Complex Natural Resource Systems: The Case of the Lagoon of Venice*, 2006, tourism management in R. Davis and G. Marvin's *Venice, the Tourist Maze: A Cultural Critique of the World's Most Touristed City*, 2005 and J. Da Mosto et al.'s *The Venice Report: Demography, Tourism, Financing and Change in Use of Buildings*, 2009, and sustainability issues in I. Musu's *Sustainable Venice: Suggestions for the Future*, 2001, and my own *Handbook of Regenerative Landscape Design*, 2008.

Given that it is now almost a decade since the last popular work by Keahey was published and, as indicated above, much specialised information has come into print in the years since, the time seems right for an update. The present work therefore encapsulates accounts of Venice's challenges and corrective measures which were covered in all the aforementioned works. Besides abstracting these foundation sources and substantially reworking a portion of the text from an earlier, technical, publication of mine, additional information for the present book has been obtained from the reports of governmental and non-governmental organisations (NGOs), newspaper articles, documentary films, other scholarly and popular books, websites, films, and many technical publications of recent provenance. Together with insights gleaned from dozens of interviews conducted while teaching and living in Venice over the years, I have distilled all this into a concise and cogent presentation – a primer as it were – about the plight of the city, its lagoon, and its people. And in order to provide context from which better to comprehend and appreciate Venice's demise, I have felt it necessary along the way to review various historical elements of her story. The novelty of the present book is in providing the first, "one-stop shopping" synthesis which gives equal weight to both the environmental and the social problems that plague Venice.

Another element of this book which is important for the provision of an overall impression of Venice's condition is the photo essays that are associated with each chapter. Although hundreds of coffee-table books about Venice have been published, these usually display beautiful images of impressive palaces, romantic canal-scapes, or happy Carnival revellers. In contrast, photographs in the present book and its affiliated website focus on physical ravages, restoration practices, and

There was an air of resignation and gloom which pervaded the city. – LISA ST. AUBIN DE TERAN, *THE PALACE*   *Venice is only an idea.* – WILLIAM RIVIERE, *BY THE GRAND*

suffocating crowds, all of which are just as indicative of modern Venice, if not even more so, than those of pretty palaces and the like. Photo essays for each of the first three chapters consist of a dozen subjects, and all essays conclude with a final image reflective of death in Venice.

In Venice, as perhaps in no other city, one is intimately exposed to nature. More than anywhere else, Venice sits – increasingly uncomfortably these days – at the interface between nature and culture, demonstrating by its very existence the false dichotomisation with which modern civilisation has regarded these two terms and the dangerous mindset that has developed as a consequence of that segregation. In reality, Venice is possibly the world's most naturalised city at the same time that its lagoon is probably one of the world's most famous culturalised pieces of nature. It is nearly impossible therefore for an observant visitor to leave without experiencing and gaining an opinion of how Venice should deal with its environmental problems. And nowhere has Venice's nature and culture been fused together more demonstrably than in the fiction that has been written about the city. The present book therefore showcases this, by containing over one hundred and fifty quotations taken from more than fifty novels that have been based in Venice. These are presented in the form of running foot-quotations which loosely parallel the material within the chapter in which they are placed. By moving back and forth between the main body of the text and these foot-quotations, the reader will be struck by both the remarkable accuracy and the incredible prescience demonstrated by the novelists.

~

There are few things more satisfying than the reading of an impassioned, well-founded, and unabashed polemic that either supports one's previously held convictions or can convince one along the way through the acumen of the presented arguments. Unfortunately for the state of the Earth's ecosystems, albeit fortunately for enjoyment of the aforementioned reading, polemics about environmental mismanagement are plentiful. One can vent frustration, for example, by reading about the loss of wildlife (*The Fallacy of Wildlife Management* by J. Livingstone)

and fisheries (*Who Killed the Grand Banks: The Untold Story Behind the Decimation of One of the World's Greatest Natural Resources* by A. Rose); the destruction of coastal wetlands (*Bayou Farewell: The Rich Life and Tragic Death of Louisiana's Cajun Coast* by M. Tidwell) and iconic landscapes (*Playing God in Yellowstone: The Destruction of America's First National Park* by A. Chase); or the proliferation of dangerous mindsets (*The Abstract Wild* by Jack Turner), to name but a few.

What these and many other similarly provocative books share is a blistering indictment against the shameless avarice, alarming ignorance, muddled thinking, and unbridled rapacity with which the planet's precious natural resources have been, and continue to be, mishandled. And in the canon of environmental mismanagements, there are few examples more egregious than that concerning Venice, the world's most beloved and most visited city.

Venice, like no other urban space, is capable of raising anger amongst the global literati. Since 19th-century art critic John Ruskin's highly influential rant, *The Stones of Venice,* there has been an established tradition of polemics written by Venetians or "Venetophiles", often in response to some other over-romanticised, preceding text. In this vein we have M. McCarthy's tourist idyll *Venice Observed* begetting P. Barbaro's resident realism *Venice Revealed*; and P. Morand's affective elitism *Venices* being counterpointed by R. Debray's caustic diatribe *Against Venice.*

I am conscious, and more than a little embarrassed, lest the present effort might be regarded as in some respects a similar set piece to *Watermark*, that wonderful book of evocative love for Venice penned by J. Brodsky. This was not the intention. Nevertheless, what follows is not by any means a happy story. Nor, I admit, is it one free of anger. But anger can be a positive force if harnessed for beneficial change, especially if it replaces ambivalence, which if left unchecked runs the risk of sliding into apathy. And there are few things that Venice can do without more than feelings of ambivalence and apathy.

In reality, the Venice of today is a very troubled city, facing extreme challenges not only from the omnipresent Adriatic, as has been bemoaned since the days of Byron and Ruskin, but also more recently from the floods of mass tourism which are just as or, if we are to believe some, even more serious. It is not idle fear-mongering to worry about and ponder Venice's long-term ability

---

CANAL Venice – the original wet dream. – ROBERT COOVER, *PINOCCHIO IN VENICE* *An unbelievable depth of having trafficked in a daydream commercially for centuries. That is*

to stay afloat, both physically and culturally. Elsewhere I have written of the prescriptive measures that might be applied to help Venice achieve that sustainable buoyancy and thus survive into the next century. Here, the emphasis is on the pessimistic back-story. The present work can thus really be regarded as more of a jeremiad than a polemic, since it is tinged more with sadness than with anger. But if it happens that anger is engendered from reading these pages, then that is also good since it may help transform apathy into action for readers, environmentalists and Venetophiles alike. Ideally, it may help provoke such Venice observers to offer support for those native Venetians – and, yes, there are still a few of those left – who refuse to give up and instead persist – either through their Sisyphean struggles to save their city, or simply through taking a stand against the barbarian-tourist hordes at the gate by continuing to live there rather than retreating to the refuge of the mainland.

Venice, with its hungry seas and crumbling structures, has always been associated with impermanence and imminent death. And, although there is no denying that this is a book seemingly written in the shadow of the grave, it must be reiterated that the danger lies in confusing pessimism with fatalism. We must strive to follow the spirit in the words of two of Venice's many famous residents: "It is late. There is much to do," said Paolo Sarpi on his deathbed; and "Ah, but a man's reach should exceed his grasp," wrote Robert Browning, who also expired in Venice. With many reaches of boldness, even brazenness, Venice *can* be saved. True, perhaps not as it once was, but certainly better than it would be otherwise.

~

I would like to thank Harvard and Ca' Foscari universities for providing me the opportunity to live and teach in Venice over the years; the many residents – students, academics, professionals and private citizens – who kindly shared with me their time and their concerns about the future of their city; R. Mi for the delightful malapropism used as the title of the final photo essay, that so perfectly sums up the governmental inability to address Venice's problems in a timely fashion, and S. Bene for some of the *acqua alta* photos.

the other thing I know about Venice. – HAROLD BRODKEY, *PROFANE FRIENDSHIP*   *Although I had never been to Venice before I had a clear picture of it in my mind.* – ANDREW

Introduction

HISTORY OF AN INVENTED CITY

VENICE, thought by Ruskin to be the greatest architectural creation on earth, is the world's most beautiful, recognisable, and beloved city. Not only is Venice the most identifiable city in the world, it is also the city that is most identified with. Amsterdam, Copenhagen, Bruges and St. Petersburg, for example, are all referred to as the "Venice of northern Europe"; Hangzhou (China), Hoi An (Vietnam), Osaka and Bangkok, among others, compete in their tourist literatures for the title of the "Venice of Asia"; as do Lowell (Massachusetts), Fort Lauderdale (Florida) and Xochimilco (Mexico) as the "Venice of North America". It seems that any place with a few canals shares the nomenclature as, for example, such well known locations as Little Venice in London or Venice Beach in Los Angeles. Further, there are over a hundred Venices in the New World: the United States has 28 and Brazil has 23, for example. Indeed, people love Venice or perhaps more accurately, the *idea* of Venice, so much, that it doesn't even matter if there is any seawater present, as witness to the Venices that exist in the most unlikely of land-locked locations such as Utah, Kansas, Missouri, Illinois, Ohio, and even Arizona. Finally, there is even a country whose very name is a derivation: Venezuela.

Venice, the "real" one, is a city that exists as much in the mind as it does in reality; it is a place of dreams and imagination into which no one ever enters as a complete stranger free of expectations and culturally acquired "memories". This is easy, at least for English language speakers, since, next to possibly New York or London, no other city has had such a magnetic attraction for writers as Venice. In one genre alone, more fictional bodies seem to pile up in Venice in dozens of crime books and films than in any other place on the planet (with the possible exception of Midsomer, Somerset in England, home to the popular *Midsomer Murders* TV series).

Sadly, Venice, the world's most copied city, may also be the world's most imperilled city. Certainly, since its conception,

Venice has always seemed to be the impossible city, defying both tides and time, the heralded symbol for the triumph of humans over nature. Today, however, all evidence points to a much bleaker future for this oft-referred to "jewel" or "Queen" of the Adriatic.

Venice is beginning to be regarded in environmental circles with the same gravitas as it has been in architectural circles since Ruskin offered it up as the ultimate archetype of form and beacon of function. Venice, many are today realising, may be the most relevant example of the tension existing between environment and development and the tradeoffs between ecosystem preservation and economic resourcism. Some have gone even further to assert that the challenge of Venice, posed as it is flush with a sea rising from climate change and subjected as it is to pressures from the travelling nouveau riche, is *the* central episode of the socio-environmental crisis of modern civilisation. The pressing question is whether Venice, the city invented in the minds of its many admirers, has itself the inventiveness to enable it to survive in the long run.

WILSON, *THE LYING TONGUE* *It was a miracle on which [he]...pondered, for the survival of this city was surely no less than a miracle.* — ARNAUD DELALANDE, *THE DANTE TRAP*

xi

Part One

SCIENCE

"Oh Venice! Oh Venice! When thy marble walls

Are level with the waters, there shall be

A cry of nation's o'er thy sunken halls

A loud lament along the sweeping sea!

If I, a northern wanderer, weep for thee."

Lord Byron, *Ode to Venice*

"Venice, lost and won,

Her Thirteen hundred years of freedom done

Sinks, like a sea-weed into whence she rose!"

Lord Byron, *Childe Harold's Pilgrimage*

Chapter 1

Acqua Alta:
Venice, Atlantis Redux

INTRODUCTION

THE dramatic conclusion in the film *Casino Royale* involves James Bond narrowly escaping from a Venetian palace as it crumbles apart around him and sinks rapidly into the Grand Canal. In a way, this is part of a long-established tradition going back to Byron and before, wherein Venice, born from the waves, is prophesied to end her days, like that of her equally illustrious predecessor Atlantis, by succumbing to the embrace of the sea and disappearing forever beneath the surface of a watery grave. Even the non-poetic are susceptible to this imagery. In December 1966, for example, a few weeks after the disastrous *acqua alta* or high water flooding event that grabbed worldwide attention, the then director general of the United Nations Educational, Scientific and Cultural Organization (UNESCO) declared that:

> "Venice is sinking into the waves. It is as if one of the most radiant stars of beauty were suddenly engulfed".

Since that time, little positive has occurred to circumvent the seeming inevitability of the city's destiny. As a result, Venice now has the dubious distinction of being included in a modern "doomsday book" with the distressing title *Disappearing World: 101 of the Earth's Most Extraordinary and Endangered Places*.

Ruskin, too, begins his remarkably influential *The Stones of Venice* with the following acceptance of Venice's aquatic extinction:

> "Her [the Garden of Eden's] successor, like her in perfection of beauty, though less in endurance of dominion, is still left for our beholding in the final period of her decline: a ghost upon the sands of the sea, so weak—so quiet—so bereft of all but her loveliness, that we might well doubt, as we watched her faint reflection in the mirage of the lagoon, which was the City, and which was the Shadow. I would endeavour to trace the lines of this image before it be for ever lost, and to record, as far as I may, the warning which seems to me to be uttered by every one of the fast-gaining waves, that beat like passing bells, against the STONES OF VENICE."

Due to the recurring threat of flooding, Venice has become emblematic of the survival of the old, the delicate, and the exotic, the city's eternal struggle against the waves being an intrinsic part of her very identity. Venice is thus the perfect allegory of decline, demonstrating the price that must be paid for egregious beauty or hubris. No wonder that Charles Dickens likened the Grand Canal to a coiled serpent wrapped around the sinning city, poised to bring about her fall from grace by pulling her into the unholy depths below.

Venice has thus become a metaphor for the death of civilisation and the triumph of elemental forces beyond human kin or control. For the last three decades, however, it has been realised that Venice's demise may really have as much or more to do with human mismanagement than from any putative whim of Nature's retribution.

> "Venice is a city beset by misfortune, suffering from severe neglect, caught in a cycle that is destroying her,"

concluded Frey and Knightly in their sentinel 1976 book, *The Death of Venice*, playfully tweaking the tile of Thomas Mann's iconic novella.

THE MARSH OF TIME

FOLLOWING millennia of fluctuating sea levels in response to global climate cycles, conditions in the Adriatic began to stabilise after the last ice age about six to three thousand years ago to form the lagoon coastline we recognise today. The Venice lagoon has a surface area of more than five hundred square kilometres, three quarters of which are open water of about a metre in depth.

Although a few Roman ruins can be found in the lagoon, colonisation really began in response to the Lombard invasions of the 6th century. Repeating the actions of the Byzantine-Romans who had established nearby Ravenna many years before, these refugees selected the location of their new settlement at a distance from the mainland precisely because it was so inconvenient. So whereas most other cities grew up because they were in good locations, the opposite is the case for Venice. It was, in McCarthy's words:

> "a foundling, floating upon the waters like Moses in his basket among the bulrushes." (McCarthy, 1963)

Once upon a time nothing was here, nothing but tidal waters, grasses, and mud. Everything was brought and constructed. – JANE ALISON, *THE MARRIAGE OF THE SEA*   This whole

The city, born out of fear and desperation, therefore came to be because the lagoon was ideal in that it was both too deep for terrestrial invasions and too shallow for maritime invasions. Venice's defensive walls became the surrounding immense "liquid plains" of the lagoon, something of obvious success given that Venice, alone of all the major Italian cities, was never successfully invaded from the time of its foundation to the 19th century.

Mudflats were consolidated from reclaimed saltmarshes by the early colonists with mats of woven reeds in a manner similar to the process used today by dwellers in the Iraqi marshlands and on Lake Titicaca. Over time, timbers were driven into the spongy soil to shore up other islets made from sediment excavated from early canal diggings. Torcello, the first large city, was formed in this way but eventually had to be abandoned in the 12th century. This occurred as a consequence of siltation resulting from rampant deforestation and increased runoff on the mainland which converted the saltmarshes to freshwater swamps, resulting in frequent malaria outbreaks. Settlers then moved further out into the lagoon to colonise a group of over a hundred small mudflat islands. Here they built the City of Venice through use of wood pilings atop of which lie cribs of wood and flexible floors to accommodate differential compaction of the sediments. Today, the famous Renaissance church of the Salute remarkably sits atop more than a hundred thousand such wood pilings. By the Middle Ages, Venice had a population of over two hundred thousand, five to ten times larger than either Rome, London, or Paris, and had become the richest city in Europe.

Venice, Ralph Waldo Emerson once quipped, was a city built by and for beavers. As has been aptly stated:

> "In the broader history of the environment, Venice was perhaps a pioneer in a common search for an equilibrium between human life and the environment's time scale". (Crouzet-Pavan, 2002)

The very existence of the lagoon has been closely dependent upon continual maintenance for more than five hundred years. The lagoon has always been threatened by alluvium deposited from the major inflowing rivers, a situation made all the worse by long-shore ocean currents which were constantly at work to fill in the open-water mouths between the outer barrier islands and thus to seal off the lagoon from the Adriatic. As the saltmarshes increased in direct consequence from the sedimentation of the lagoon, so did the fears that Venice would lose its protective moat as well as be deprived of peaceful shipping, its mercantile lifeblood.

The Venice lagoon has always existed in a fragile, dynamic equilibrium upon which humans have acted, no more dramatically so than through the bold series of modifications involved in diverting the inflowing rivers between the 13th and 16th centuries. Fears that the lagoon would disappear remained a constant concern in Venetian life. A huge debate raged in the mid-19th century, for example, when the occupying Austrians diverted the Brenta River back into the lagoon to reduce the flooding threat to upstream Padua, which by that time had become a more valuable asset.

The construction of the causeway by the Austrians in 1846 – famously referred to by Venetians in a mocking spirit of their one-time, renowned arrogance, as finally enabling the mainland of Europe to be linked to Venice for the former's benefit – changed the city from being a cluster of islands to a peninsula. This has resulted in the "devastation" of the western end of the historic city, through providing a "bridgehead [allowing] the twentieth century's assault on Venice to accelerate rapidly" (Lauritzen, 1986). Today, one of the largest car parks in the whole of Europe dominates the area, resembling as it does one of those huge alien spaceships hovering ominously over other famous historic sites in the movie *Independence Day*. Indeed, there are those with Venetian familial roots going back centuries, such as television personality F. Da Mosto, who believe that it was the construction of this causeway, identified by some today as the modern Bridge of Sighs, that was ultimately responsible for bringing about the end of the great Republic of Venice, not, as is commonly supposed, Napoleon's invasion of the half century before.

THREATS AND CHALLENGES

That Sinking Feeling

WHEN undertaking repairs to the Ducal Palace in 1810, French engineers were surprised to find that the "missing bases" of the

---

city is set on a base of wooden piles, thousands of them, and they are all slowly rotting. – EDWARD CHARLES, *DAUGHTERS OF THE DOGE*   *Venice is not known for its enthusiasm*

lower columnar arcade were actually 38 cm below ground level. The arcade and palace had obviously sunk over the five centuries since its construction and this had been compensated for by a raising of the surrounding pavement. Subsequent archeological work in Venice and the lagoon has confirmed these early findings. Evidence from Torcello shows that residents struggled to stay ahead of the general subsidence of their island by repeated building, one layer upon another. Elsewhere, a Roman walkway now lies 2 metres below the current sea level. Even much more recent structures, such as the crypt of St. Mark's Basilica in the historic centre of the Venice, are now actually 20 cm below present-day sea level. What is going on? Is Venice *really* sinking?

Venice sits on top of a squashed saltmarsh and mudflats which are themselves situated above a layer of several kilometres of river sediments that are all slowly being compacted. Venice, as it presses down, squeezes out water and the sediments become thinner and thinner. The result has been a subsidence of several millimetres per year. The situation has been exacerbated by the diversion of the major rivers away from the lagoon, resulting in centuries with no ongoing supply of alluvium to counter the natural subsidence. And if that was not serious enough, humans have greatly worsened the problem.

Measurements made in Venice indicate a subsidence rate over the last century of 2.6 mm per year. This is a rate that is more than twice that of the average for other locations around the upper Adriatic. Venice actually subsided 17 cm between 1908 and 1961 as a direct result of industrial operations on the mainland. In contrast to the Venetian tradition of decades of debate ensuing before decisions are made and actions taken, plans for constructing the mainland port of Marghera were accepted just five days after submission in the 1920s due to being tied to the wishes of the nascent fascist party. Ultimately this resulted, by 1969, in the five tap-wells at the port withdrawing an incredible 1,500,000 litres an hour of groundwater to sustain the industries located there. This meant that in its battle against the sea, Venice sank an additional 10 cm (relative to the nearby city of Trieste) during the eighty years the pumps were in operation. In other words, had it not been for Porto Marghera, it would have taken Venice until 2050 to reach its current elevation. Humans therefore accelerated the natural rate of subsidence by about a century.

An intriguing confirmation of Venice's subsidence comes from study of the meticulous 18th-century paintings of Canaletto and his studio which were made using a camera obscura. This was a technique in which a reflected image of a scene is displayed in a dark box, enabling it to be copied as a painting. This enabled an explicitly accurate representation of a scene to be made about a century before the invention of photography. A careful comparison of such paintings with modern photographs taken from the exact locations showed that today the everyday wetting of the same buildings and consequent buildup of attached algae are 60 cm higher than in Canaletto's time. This represents an average rate of subsidence of 2 mm per year. Another ancillary demonstration of differential changes in the elevation of the City of Venice compared to the sea comes from the fact that it is now impossible to paddle replicas of historic gondolas, which have much higher bow-peaks than do today's boats, about much of the city since they can no longer fit under the bridges.

The bottom line is that tides of one metre, which would have flooded less than five per cent of the historic city a century ago, now submerge almost forty per cent of the area. And, if we are to believe predictions from climate change models, things are certainly not going to improve in the near future.

Irreconcilable Differences

VENETIANS have always had "an unusually intimate relationship with the sea" (McCourry, 2005). In the halcyon days of the Republic, the city would honour their maritime dependency by reaffirming its vows with the sea through an elaborately staged marriage ceremony. Every year the *Doge* would sail out into the Lido basin on a magnificent barge accompanied by a flotilla of boats and engage in a matrimonial ritual culminating with him casting a gold wedding ring into the waters. Today, all is not well in this domestic relationship and it seems that, with construction of the flood barriers (described below), Venice and her spouse are finally on their way to divorce.

Images of a precarious Venice surrounded by a malevolent sea have long been commonplace. Recently, for example, the cover of *The Science of Saving Venice* shows a frothy, white-capped lagoon battering against the fragile looking Molo *fondementa* or embankment in the background; and for the alarmingly titled

for nature; indeed, its entire history has been dominated by the need to keep the elements from interfering in its business. – IAIN PEARS, *THE TITIAN CONSPIRACY* 'Venice is a dead

Venice Against the Sea: A City Besieged, the artsy cover shows a distant view of the tiny city sitting all alone in a vast maelstrom that is actually superimposed on top of several of the buildings. Such futuristic imaginings have been taken to the extreme in the Nova documentary *Sinking City of Venice,* in which, in a scene reminiscent of those from the movie *Waterworld,* a computer-simulation shows Venice of several hundred years hence, with only the top third of the Campanile tower and the dome of St. Mark's Basilica poking out from above the churning waves.

Today, global warming is simply making a bad situation worse. Sea levels around the world are predicted to rise anywhere from 8 to 88 cm by the start of the next century. Due to being built flush with the water, Venice is the first major Western city to face sea level rise as a consequence of climate warming. The present conditions in Venice are therefore a presage for what the future might bring elsewhere. In this regard, believers in techno-fix solutions like the producers of the documentaries *Go Deep: Saving Venice* and *Venice: Code Red* believe that the city functions as an engineering testing ground in the battle against climate warming. In other words,

> "In order to save its past, Venice is a key to the future of the world's coastal cities." (History Channel, 2009)

Others are not so optimistic. The sentence

> "sea level rise is like a rope that is tightening around the neck of Venice" (Discovery Channel, 2008)

though over-dramatic, may however also be apt. The final words of the A & E documentary about the fate of Venice, spoken in baritone gravitas by none other than Chief Science Officer Mr. Spock himself (aka Leonard Nimoy), intones that "eventually the city that seemed impossible to build could be impossible to save." In other words, the depressing reality may be that in order to save Venice, the climate of the planet itself must first be saved. And of course, as witness to the disappointing results from the recent world climate congress in Copenhagen, we seem remarkably unable to accomplish the latter task.

Waterscape Alterations

VENICE'S lagoon has long been overshadowed by the city's architecture. Actually the lagoon contains the largest (70 km^2) wetland in Italy and one of the most important in the entire Mediterranean in terms of avian biodiversity and abundance, and would be famous today even had its accompanying, dreamlike city never existed. The rest of the lagoon is composed of 90 km^2 of closed-off fish farms, fifty islands, and over 1,500 km of canals.

Although frequently ignored in discussions about the fate of the Venice lagoon, it is important to recognise that such bodies of water are really transitional landforms. All lagoons represent a delicate balance between competing terrestrial and marine influences and are therefore only temporary ecotones or ecological boundaries on their way to evolving into either land or sea. To regard lagoons as static entities stuck in some sort of medial permanence is to deny the inevitable processes of geological and ecological succession. And in a sense, this is exactly what Venice's environmental history has really been about: continually altering the lagoon by attempting to bend time's arrow back upon itself to some perceived optimal but ultimately unstable single moment. Although it may be a hard reality to face for some, the Venice lagoon of today is simply not "natural" in any true sense of the word; instead, it is an artificial construct which would have long since disappeared had it not been for deliberate and repeated human interventions.

The opposing and dynamic tension between land and water has always been the key element determining the fate of Venice and the other island settlements in the lagoon. For example, an 18th-century engraving representing the lagoon (which was appropriately used as the cover image for the book *The Venice Lagoon Ecosystem: Impacts and Interactions Between Land and Sea*) shows a pitched battle being fought between terrestrial and aquatic gods portrayed as two giant wrestlers towering over their prize, the City of Venice. Others have used the military metaphor such as Keahey in his book title and C. Scearce:

> "for centuries the inhabitants of Venice have waged war on natural processes of change, in order to shape the [lagoon] city to their needs." (Scearce, 2006)

Lagoon ecosystems in general, and that of Venice in particular, through being in a characteristic state of disequilibrium, are vulnerable to change. Since the diversion of the major rivers in

old city. That's why you like it.' – EDWARD SKLEPOWICH, *THE VEILS OF VENICE   Her eyes watered and he wondered if Torcello's sad atmosphere had reminded her of memories*

7

the 14th century, the Venice lagoon has been an artificial body of water. And from the mid-19th century onwards, the lagoon has completely lost its ecological balance due to a host of morphological and hydrological modifications. These include construction of jetties at the mouths of the three inlets from the Adriatic, compartmentalisation of the lagoon into isolated fish-rearing enclosures, conversion of marshes, mudflats and shoals into dry land, dredging of two large shipping channels between the barrier islands, and the accelerated subsidence due to industrial groundwater extraction.

At one time the lagoon was a river delta, a swampy marsh into which three major rivers and many tributaries emptied and dropped their loads of sediment. Following diversion of the rivers, the channel openings to the Adriatic were modified to increase tidal currents to sweep the lagoon free of accumulated sediment. In the 19th century, for example, shifting currents from a newly built deepwater channel actually caused an entire island to disappear, forcing its residents to be relocated. For the last fifty years, the openings to the Adriatic between the barrier islands have been repeatedly enlarged and increasingly deepened in order to permit passage of progressively larger ships into the lagoon. The reason for this has been economic. Today, the Venice Port Authority approaches one billion euros in business a year (about a third of the amount for New York City). Employing 18,000 people and handling 29 million tons of goods annually, it is a substantial contributor to the regional economy. With more than 150 ships mooring at any one time, the port is the second largest in Italy and the eighth most important in the world. The deepwater channels have allowed for an increase of about ten per cent in the amount of water entering during high tides. This has led to a doubling in the total volume of water present in the lagoon compared to the 19th century. The deepening and narrowing of the shipping entrances and canals has also seriously interfered with sediment dynamics. As well, mussel fishermen continue to use illegal fishing methods that scour the lagoon floor. The combined result is that more than a million cubic metres of sediment is washed away from the lagoon every year. As a result, the lagoon is now depleted – some researchers use the word "starved" – of sediment to the extent that vast areas exist of barren underwater desert.

Fish farming and industrial and urban infill have reduced the overall size of the lagoon. Inhibited water movement associated with fishing weirs and holding pens was recognised as early as the 13th century, but issues of food production security always trumped hydrodynamic concerns. The situation is more serious today given that the one-time netted fish enclosures have been replaced with earthen dam enclosures that have further reduced the lagoon's area and more severely interfered with the natural circulation patterns.

Hydrodynamic alterations have also occurred in the City of Venice itself. At one time the city relied upon the twice daily flushing of tides to remove waste discharged directly into its canals. However, starting in the 19th century, the "pedestrianisation" of Venice for tax benefits and to increase tourist access has meant that this "dilution is the solution to pollution" cleansing system no longer operates efficiently. Of the original network of canals, 40 km were filled in between 1815 and 1889. The conversion of Venice's original canals into pedestrian thoroughfares has not only significantly altered water flows but also substantially changed the very makeup of the city. The Venice of today is, as it never has been before, primarily a walking city.

WETLANDS OF MASS DESTRUCTION

THE saltmarshes around Venice have been under attack since the first settlers entered the lagoon. Over the centuries various proposals have been made to fill in most of the lagoon for agriculture (as was done elsewhere in Italy). The area of mainland facing the lagoon was once a verdant wetland paradise. The poet Dante described the area as being so full of vegetation that he had difficulty walking. Later, the playwright Goldini waxed about the garden villas that were surrounded by bird-filled wetlands. Today, however, the mainland coast is a sprawlscape of suburban concrete and factories.

Porto Marghera, which sits on the mainland directly across from the historic city, began after World War I with the goal of rescuing Venice from its post-Napoleonic poverty by recapturing its past glory as a centre of maritime commerce. The initial development of the first industrial zones was brought about at the

better left forgotten. 'This place is all gone,' she muttered. – STEVE BERRY, THE VENETIAN BETRAYAL   Through just a few inches of brackish water he could make it out: fine ancient

expense of destroying 4,000 acres of wetlands. This was followed in the 1960s by another loss of 10,000 acres of wetlands to enable development of one of the largest petrochemical complexes in Europe. Today, the site is also one of the most energy-intensive industrial areas in the entire European Union, with fully ten coal, oil, and gas-fired power plants feeding into the national grid. The overall project has always had its critics, many of whom would wholeheartedly agree with G. Pertot's criticism that it was and is

"an extraordinary, unbelievable intrusion on the Venetian landscape" (Pertot, 2004)

that was developed under the guise of a hare-brained pretension that the city could still be a major economic player.

And then there is the neighbouring dormitory city of Mestre, an unattractive yet functional jumble of apartment buildings and offices, many built on the sites of drained saltmarshes. Mestre's population grew five-fold in the second and third decades of the 20th century due to its proximity to the industrial juggernaut of Porto Marghera. This enabled it to sprawl

"like an oil slick as a consequence of the indiscriminate parceling out of land for building development and absence of any concern for urban and environmental values." (Pertot, 2004)

As a result of these and other developments, from half to three-quarters of the surface area of saltmarshes have thus been lost from the Venice lagoon since the start of the 19th century. Today, only 9.900 acres of saltmarsh remain. Not only has this meant a twenty per cent loss in plant biodiversity and a fifty per cent decrease in bird species in the lagoon, but the loss of wetlands, in combination with the scouring away of sediments and the increasingly high waters, means that the Venice lagoon is functioning today more like an open-water marine environment than a wetland.

There is little cause for optimism. The lagoon is now so starved of sediments that the saltmarshes can no longer establish themselves naturally. The saltmarshes continue to be chipped away by increased wave action in the enlarged boat channels. As a result, the bleak conclusion is that all the wetlands could be gone within another century.

THE MOST BEAUTIFUL TOILET IN THE WORLD

AT one time it was against the law to damage Venice's waters. In the Museo Correr one can see a wall plaque from 1553 with the following inscription from the *Magistrato alle Acque*, the Republic organization charged with protecting the lagoon:

The city of the Venetians
With the aid of Divine Providence
Was founded on water
Enclosed by water
Defended by water instead of walls
Whoever in any way does damage to the public waters
Shall be declared an enemy of the State
And shall not deserve less punishment
Than he who breaches the sacred walls of the State
The edict is valid forevermore.

Today it seems those who are responsible for the deteriorating environmental quality of Venice are much more likely to be promoted than punished. "Forevermore" seems to have become "nevermore".

Daniel Craig, as super-spy James Bond, is seen swimming in the crystal clear water of the Grand Canal in the penultimate scene of *Casino Royale* as he struggles to escape from the rapidly sinking palace. This might be the most inaccurate image in the entire film, given the opacity of Venice's canal water due to nutrient pollution and consequent algal growth in addition to the constant resuspension of contaminated sediments. The reality, as known by native Venetians who shake their heads in disbelief at the antics of some egregiously naïve tourists, is that one should never swim in Venice's canals…*ever*.

Venice's canals used to be routinely drained and dredged. Since the 1960s, however, regular maintenance has stopped, which has resulted in an accumulation of sediments laden with heavy metals, mercury, and fecal coloform bacteria. By the mid-1990s, people were referring to the Grand Canal as "a bath of poison". One wonders that if 007 had known that bioassays of sediment elutriates taken from inner city canals showed them to be highly toxic, he might have hesitated before plunging into the water in attempt to rescue his love interest. The film ends before

bricks in a herringbone pattern, clear water lapping over them gently. It was the floor of a villa two thousand years old. — JANE ALISON, *THE MARRIAGE OF THE SEA* Venice had

we learn the true implications of Bond's immersion. In the real world, one might expect him to have to visit a health clinic upon arrival back home and to be the recipient of antibiotic-filled needles to mitigate the deleterious effects of contact with Venetian waters.

Once it was possible to be able to look northward from the top of a Venetian campanile on almost any day of the year and be able to see the Dolomite Mountains, sixty kilometres away. For almost a century, however, when tourists arrive at Venice by land, the most overt sign of pollution they will notice will be the dark clouds emitted from the forest of smokestacks in the mainland industrial complex of Porto Marghera. The contrast to the white shimmering beauty glimpsed on the nearby islands of Venice is so glaring that it is impossible to resist calling to mind William Blake's still-powerful words of reference to "the dark, Satanic mills" of the industrialised English Midlands. Emissions of sulphur dioxide from these petrochemical stacks create clouds of polluted air that hang over the entire lagoon and contribute to corroding Ruskin's hallowed "stones of Venice" at the same time, ironically, as contributing to the beautiful red sunsets that are so admired by tourists.

Water contamination, although more subtle, is thought by many to be a more serious concern. Only in the aftermath of the famous 1966 flood (see below) did the magnitude of the pollution problem become realised by the international community. Hundreds of tourists were stunned as they stared at the disgusting scum that coated the entire city as the high water or *acqua alta* receded. Today, industrial discharge and leaching from associated toxic waste sites in Porto Marghera, urban runoff from Mestre, intensive agricultural drainage from the watershed, and a lack of adequate sewage treatment in Venice itself, all contribute to the contaminated cocktail of the lagoon waters.

For centuries, visitors to Venice have been shocked by the degree of pollution. "Do not piss here" signs were once prevalent and local women used to wear *chopines*, 60-cm high clogs to raise them out of the filth of the *calli* or pedestrian walkways. Sanitation waste was simply dumped straight out of windows, just as it always had been in thousands of medieval cities. But in Venice's case this practice continued long after those other cities had moved to collect and treat their sewage. Today's tourists would be shocked, as they flush their hotel toilets, to learn that the "modern" city still has no sewage network and that "huge quantities of human waste flow directly into the lagoon's waters every day" (PRESUD, 2004) where it combines with untreated waste from restaurants. Together with the odour of diesel exhaust from the hundreds of service boasts, the smell of rank sewage exposed at low tides has "become a regular summer feature" of the modern Venice experience.

At more than 2,000 hectares in size, Porto Marghera is the largest contaminated brownfield site in all of Italy. Over the years, the production at this location of seventy per cent of the oil-derived chemicals in Italy has resulted in thousands of tons of heavy metals and organochlorines being dumped into the lagoon. Here, the consequent toxic sludge is repeatedly resuspended in the shallow waters by wind and boat wakes. None of this should be surprising since it has been by deliberate design and not through happenstance. Indeed, the environmental damage wrought by the port was officially recognised as far back as the 1950s, as shown in one early planning document:

> "To be located in the industrial zone are mainly those plants that discharge smoke, dust or emissions harmful to human life into the atmosphere, that discharge poisonous substances into the water, and that produce vibrations and noise." (Cited in Pertot, 2004)

The plan to promote the area as a pollution haven has succeeded to such a degree that today the port contains factories, power stations, processing plants for pyrite, steelworks, in addition to one of the Europe's largest petrochemical complexes. That all this is located within a few kilometres of the world's most beautiful and possibly fragile city, beggars belief.

More than a million people live within the 1,877 km² drainage basin of the Venice lagoon. Eighty per cent of the land surface area comprises agricultural fields which, until the 1960s, had relied upon antiquated farming methods. Fertiliser in the runoff from these fields mixed with detergents disposed directly into the Venetian canals contributed to a serious problem of eutrophication or over-enrichment in the lagoon during in the 1980s and 1990s. This resulted in the proliferation of noxious accumulations of macroalgae which were able to out-compete

endured so many trials, he thought. It was unquestionably the jewel of the Mediterranean, but it was a fragile jewel, constantly under threat from tides and storms. — ARNAUD DELALANDE, *THE DANTE TRAP*

10

the small planktonic algae for uptake of the abundant nutrients. Upon death, these macroalgae sank to the bottom and began to decompose, which led to the depletion of oxygen in the water. The net result was the mortality of millions of bottom-living organisms and the deaths of thousands of fish.

IMPLICATIONS

Taking the Waters

VENICE is not so much a city built on the water as it is a city built in it. Even in the absence of all exacerbating human influences, Venice would still be predisposed to flood due to its location at the northern cul-de-sac of the Adriatic combined with regional weather patterns and tides. The *sirocco* winds from the southwest blow along the complete length of the Adriatic. Because the basin becomes progressively shallower toward the north, water essentially piles up at the top end. Further, the funnel-like shape of the coastline at this location serves to focus the wind and waves toward Venice. And if this were not enough, the *bora* winds from the east push all this accumulated water deep into the lagoon. The consequent seiche* or internal wave generates water levels that are 60 cm higher in the lagoon compared to the Adriatic. The result of all this is that Venice can experience average high tides that are generally 1.2 m above sea level.

As the world learned with respect to hurricane-ravaged New Orleans, Venice's vulnerability to flooding has also been increased due to the cumulative destruction of the surrounding wetlands. With the removal of the complex network of dendritic creeks and the smoothing out of the rough bottom of the lagoon, the ability of the lagoon wetlands to absorb and dampen tidal surges has diminished. Due primarily to subsidence, and also perhaps to erosion of wetlands, the volume of water in the lagoon has more than doubled from what it was a century ago.

During November 3 and 4, 1966, a major storm raged across northern Italy that severely damaged Florence and left more than a hundred people dead across the region. In Venice, water levels peaked at 2 m over sea level and the entire city was flooded for only the fifth time in its history. The inundation lasted for over 15 hours and the city was without electricity or phone services for more than a week. The islands of Pellestrina and San Erasmo were completely submerged. Although Venice had experienced several serious floods before, the 1966 flood was the first time that water had risen high enough to inundate the interior of St. Mark's Basilica. The flood, by bringing Venice's destiny to international attention, became a major wakeup call for those concerned about the city's future. UNESCO's report on the incident, written a decade later, concluded that the flood had "revealed the negligence, lack of care and of skill from which Venice suffered and the scourge of industrial development".

Water was not the only problem, however. Just as in New Orleans following Hurricane Katrina, there was a massive leaking of chemical waste that left behind a black, contaminated coating that blanketed every surface. Once the waters receded and the grime was removed, almost all of Venice's buildings were in need of substantial structural repair. Nevertheless, in spite of the hullabaloo raised in the foreign press following the flood, the fact that there were no resulting deaths and that no art masterpieces had been damaged led Venetians, once the shock of the magnitude of the event had subsided, to resume their characteristically lackadaisical attitude toward flooding. In point of fact, Venetians have always been resigned to *acqua alta* events, for the most part regarding them as being much a part of the life in their city as do Montrealers, for example, snow blizzards.

> "What's the problem? The water goes up; it goes down. No one is hurt. This has been happening for centuries and we're still here,"

reported one native. Even the mayor seemed unfazed, quipping that

> "Venetians have been getting their feet wet for centuries."

The flood in 1966 was an extreme example of what has become an all too common phenomenon in modern Venice. What were once exceptional events have become more and more commonplace. During the decade of the 1920s, for example, 385 *acque alte* were recorded. The incidence has increased progressively

---

* A large-scale, internal wave of sloshing water (as in a bathtub when one immerses)

Like any pampered spouse, Venice had also been neglecting her partner, the sea. – MICHELLE LOVRIC, *THE FLOATING BOOK*   *The ice was melting at the pole and adds daily to the*

11

throughout the century culminating in there being 2,464 such events during the 1990s. The magnitude of the flooding has become worse, such that eight of the ten most extreme floods of the last century have been concentrated within the last forty years. St Mark's Square, which used to be flooded about ten times a year a century ago, about twenty times a year in the middle of the 20th century, now floods up to one hundred times per annum. In fact, the situation has become so severe that water now invades the atrium of St. Mark's 150 to 180 times a year, including often every day during the winter. And what was once unheard of, summertime flooding, now occurs each year. Venice, it now seems, is finally at the point of losing its long battle against the sea.

In addition to the structural damages detailed below, Venice's increased rate of flooding has enormous economic implications in terms of time lost due to reduced or interrupted mobility, interruption of revenue activities and services, and damage to goods stored in warehouses and shops. And then there are the cleanups that inevitably must follow every *acqua alta* for:

> "all around [there is] the stink, the miasmas, because the water smells of latrines and of sulphur, it washes out the sewers, insinuates itself into the tight places and floats out the filth. There also bob along the *calli* the sacks of garbage." (Cited in Davis and Marvin, 2004)

Finally, flooding also has serious health and safety implications. The worst possible time to suffer either a heart attack or a home fire is during an *acqua alta* because water ambulances and fire-boats are frequently unable to squeeze underneath the bridges at such times. Denied rescue, both victims and homes succumb. But despite all this, the city fatalistically keeps muddling along, tempers soothed by the nearly five kilometres of *passerelle* or raised duckboard walkways that are deployed at times of *acqua alta* to enable people to continue on with their daily lives, elevated a few centimetres above the noxious water.

HUMPTY DUMPTY'S BANE

IN 2007, a piece of façade fell off the Ducal Palace, the building championed by Ruskin to be the single most important architectural structure in the world, Now, whereas the odd piece of tumbling statuary might otherwise go unreported – except of course by writers such as J. Berendt who titled his bestselling book, *City of Falling Angels*, about this – when a piece of white marble from the one-time residence of the *Doge* right in the heart of St Mark's Square flakes off like an iceberg from a calving glacier, people *do* begin to notice. But none of this is new of course.

For many centuries, Venice, the greatest work of art in the Western world, has proudly displayed its monumental structures in bold defiance of Nature, flaunting the absurd improbability of its very existence. Since the fall of the Republic, however, Venice has instead been identified as the most obvious example of the inevitability of Time's ruin, a Miss Havisham sort of place where all the buildings bear scars of two centuries of neglect and gradual deterioration. Decay and even death, it must be admitted, is undeniably an essential element in Venice's appeal. And many are the visitors who have purposely sought out the city in order to relish in its slow and ravishing obliteration. Spurred on by the gloomy writings of Ruskin ("The rate at which Venice is going is about that of a lump of sugar in hot tea") and others, Victorians rushed to see the city before it disappeared forever. By linking the ruins wrought by time to the decadence of its human residents such as Casanova, many came to regard Venice as a modern-day Sodom and Gomorrah whose fate seemed justified as retribution for its many sins (most notably that of pride). Venice thus became both the quintessence of stricken beauty and the archetype of exquisite corruption.

The "romance of decay" and the ensuing cult of crumbling decadence have been fed in recent years by the increasing floods that have ravaged the city. Venice's tides have some of the highest salt and oxygen concentrations recorded in the world. Because the city is largely composed of brick, this becomes a recipe for disaster. Due to increased tidal exchange with the Adriatic, canals now have the same salinity as the sea. The corrosive damage to bricks has consequently increased markedly.

Venice floods as a result of the water pouring in from the lagoon but also due to water bubbling out of drains and sewers all over the city. Dissolved salt is deposited on bricks, and because of the high relative humidity, it crystallises and bores its way into the structures. As it expands upon drying, salt causes bricks to

level of the sea. And each day Venice sinks by just so much of a fraction. – SALLY VICKERS, *MISS GARNET'S ANGEL* *All night the canal water rustled like brown silk outside my*

crumble and disintegrate. Remarkably, during and after an *acqua alta*, capillary action can actually pull the salt two to five metres upward in the porous bricks. The salt water and mixture of contaminants essentially turns the bricks to pulp. Soon the stucco begins to peel off which in turn exposes the ends of the wooden floor planks and iron rods to the corrosive atmosphere which in turn adds to the overall instability.

Exceptionally low tides, which are also a recurring feature of the hydrologically altered lagoon, are thought by some to be even more harmful than *acqua alta*. Once exposed to the air, the heads of wood pilings begin to decompose, undermining the foundations. Many of Venice's historic buildings are ingeniously constructed with walls that can move up and down somewhat independently of one another in response to localised subsidence, the whole lot being held together by iron tie-rods. Upwardly mobile salt and dampness, however, corrodes these retaining structures such that in extreme cases the entire building can become unstable and deformed, occasionally causing the outer walls to buckle and collapse outright into the hungry water.

The whole process of deterioration is exacerbated by the wash from boat propellers which generates churning waves that can add another 40 to 50 cm to the height of water exposure to buildings. The motorised *vaporetto* or water bus service began in the 1880s and the danger of boat wash was first mentioned in the late 1950s. Over the years, there have been 189 attempts by local government authorities to address the problem of propeller wash caused by the four thousand motorboats (including eight hundred water taxis) that churn up the Grand Canal each day. And it is not only buildings that are damaged in this way. Estimates are that the near constant buffeting to which gondolas are subjected to results in decreases in their life expectancy from forty to ten years.

The structural dangers posed by corrosive seawater have always been recognised by Venetians. For this reason, building foundations were composed of impervious Istrian marble, characterised by a low porosity. For centuries, this worked perfectly. Today, however, the combined actions of propeller wash and boat wakes added to land subsidence and sea level rise has meant that the higher porosity bricks which originally would have always been high and dry, are now repeatedly, and in some cases constantly, exposed to water and salt. Added to all this is the sucking action of the rebounding waves as they return to the canal, which can actually physically tear apart the salt-damaged bricks. Propeller wash and boat wakes also resuspend sediments from the bottom of the canals. This material blocks the external drains and antiquated sewage outlets from the buildings, causing them to rupture internally and damage building foundations and canal walls. Because of all these factors, many have come to believe that motorboats may actually be causing more structural damage to Venice's architectural heritage than *acqua alta*.

Reparations to Venice's buildings as a consequence of water-related damage are estimated to cost between three to five million euros a year. This does not cover indirect costs such as lost revenue, and expenses resulting from being closed for business and having to repeatedly clean out the accumulated grime.

And then, as if the effects of water and salt were not enough, there is the structural damage caused by the infamous "flying rats" of Venice. There are now close to two hundred thousand pigeons occupying the historic centre where, until recently, they were supported by food handouts from gullible tourists. The problem is that, once the tourists depart in the evening, these filth-ridden birds are forced to scavenge and peck away at scraps that have blown and settled into crevices between the brittle bricks, thus adding to the overall deterioration.

---

hotel window, licking the pavement with its high tide. – LISA ST. AUBIN DE TERAN, *THE PALACE*   *Even at low tide the steps get covered now.* – JOSEPH KANON, *ALIBI*

THE SCABS OF VENICE

HYDROLOGICAL MODIFICATIONS: The Venice lagoon has undergone extensive hydrological modifications including widespread partitioning into aquaculture allotments (1), huge openings gouged out between the barrier islands to enable freighter traffic (2), and bisection from an elaborate network of channels for boats (3, 4). The City itself is also not immune in that hundreds of one-time canals have been filled in and converted into pedestrian ways (5), often discernible by the presence of colonnaded vestigia (6) as well as planted trees as if in a mainland city (7, 8).

Colour versions of all uncropped images can be accessed at **www.libripublishing.co.uk/veniceland**.

Torment by water. – LOUIS BEGLEY, *MISTLER'S EXIT   To the south was the lagoon where the waters waited to swallow Venice up. It was too awful to think about. –* SALLY VICKERS, *MISS*

GARNET'S ANGEL   There is a certain melancholy beauty to the way she [Venice] flirts with the sea, like a lovely woman lifting up her decorated skirts – sometimes not far enough – to miss

SALTMARSHES: The Venice lagoon has lost from half to three-quarters of its saltmarshes as a result of coastal development infill (1, 2), reclamation drainage trenching (3), and the constant battering of violent waves from boat wakes (4, 5) that erode the edges of the marsh (6), causing sections, first minor (7) then spreading to massive (8), to slough off and disappear.

the rising tides. — SARAH DUNANT, *IN THE COMPANY OF THE COURTESAN*   *The strange façade of the Basilica San Marco was reflected like a mirage of a fairy-tale castle in black puddles*

along the pavement. Venice was sinking, everyone knew that, it was only a question of time. — ROBERT GIRADI, *VAPORETTO 13*   *Venice has its feet forever in water.* — SALLY VICKERS, *MISS*

NATURE: Although nature seems elusive in Venice it does poke out here and there sometimes in the seemingly most unlikely of places such as the saltmarshes beside the airport (1) or the woods of the old Jewish cemetery on the Lido (2). Away from the tourist boats and crowds it is still possible to discover natural areas of the lagoon such as the biologically rich mudflats (3) and the beautifully flowered saltmarshes (4). Nearby Venice is the blissfully peaceful island of San Erasmo where it is possible to walk alongside expanses of littoral saltmarshes (5) as well as agricultural fields and treed copses that mesh with pastoral canals (6, 7).

GARNET'S ANGEL   The [water] taxi picked up speed and rapidly plunged into the misted lagoon that sheltered the slowly drowning city. – DAVID ADAMS CLEVELAND, WITH A GEMLIKE

FACTORIES: Vast complexes of smokestacks and their complicated networks of spaghetti–like piping are observed as one approaches Venice from the mainland by train, tour bus, or car (1, 2, 3, 4, 5, 6). The incongruity of this juxtaposition is amplified on those evenings when one can gaze westward past the palaces and churches in the historic centre and see flames shooting up a hundred metres into the night sky from the venting stacks in a scene calling to mind Blake's expression about the "dark, Satanic mills".

FLAME   The driver propelled his boat along at what seemed like an impossibly reckless speed, weaving in and out of the traffic, and aimed it at the side of the canal. — IAIN PEARS, *THE TITIAN*

INDUSTRIAL PORT: Located just a few kilometres from historic and picturesque Venice is its mainland industrial port, Porto Marghera, one of the busiest in all of Italy. Here hundreds of large ships enter (1) for repair yard servicing (2) or to load or unload containers (3, 4, 5). Also products from the nearby factories (many petrochemical) held in small (6), medium (7) or large (8) storage units are loaded onto ships (9) as are enormous piles of scrap and other industrial waste (10, 11, 12). The port is really a vast brownfield site that contains many derelict buildings (13) and contaminated landfills (14).

CONSPIRACY A water taxi roared away from the neighbouring jetty, sending up a swell. – DAVID HEWSON, LUCIFER'S SHADOW The V-shaped ripple of the gondola clucked and sucked

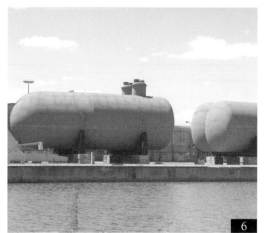

at the walls of crumbling tenements. – L. P. HARTLEY, *SIMONETTA PERKINS*   *Boats of all sorts ploughed up and down the Grand Canal, making waves. –* IAIN PEARS, *THE TITIAN*

POLLUTION: Even though alteration of antiquated agricultural practices has substantially reduced the input of nutrients into the lagoon, it is still possible to find blooms of noxious algae in many areas of restricted circulation (1, 2). Whereas some of the larger hotels and apartment blocks have their sewage pumped into "honey boats" for removal and eventual disposal (3), untreated waste from many other dwellings is still dumped into the canals causing proliferation of attached algae which can often stink during low summer tides (4). The lovely shimmering reflections that one sees everywhere in Venice's canals (5) are caused by light refracting on vast quantities of gasoline and diesel fuel leaking into the water from thousands of boats.

CONSPIRACY *He cut a swath through the water and sent swells heading toward the other low-rising craft.* — STEVE BERRY, *THE VENETIAN BETRAYAL* *Even the ducks don't come anymore;*

SUBSIDENCE: The degree of Venice's subsidence can best be observed on the island of Torcello where, standing on the present ground level that has been repeatedly built upon over the centuries, one can look down on the foundations of the oldest buildings in the lagoon (1, 2). Differential settling occurs throughout the lagoon, discernible by leaning church campaniles such as the one on the island of Burano (3) or any number of twisty palaces in Venice, as for example the Ca' da Mosto, one of the oldest buildings on the Grand Canal (4). Even more recent palaces display tilting balconies and misaligned doors (5). The front steps of palaces such as the Ca' da Mosto (6) and others of similar antiquity (7) are now completely submerged even during low tides.

the motorboats have driven them away. – BARRY UNSWORTH, *STONE VIRGIN*   *'The scheming little bastards filled in the lagoon for more industry and airports and gouged out channels for*

tankers and that changed the very tides, eroding all the foundations.' – ROBERT COOVER, *PINOCCHIO IN VENICE*   *'From today I don't smuggle out the rubbish. I put it in the canal like*

FLOODS: High water or *acqua alta* events have become commonplace in today's Venice. The rising water pours in from the lagoon overflowing the canals and making walking dangerous (1). Although the angled light along flooded arcades might look attractive (2) few would want to venture into other more ominous looking inundated areas (3). Flooding of inland areas also occurs when water seeps up through the ground (4) or erupts upward from drain-holes (5: central left of photo). As flood waters converge more and more of the city becomes submerged (6, 7). Meanwhile, building pumps keep sending the intruding water back out into the streets (8, 9) from where of course it flows back inside again.

the other people, when there's no police boat coming up.' — MURIEL SPARK, *TERRITORIAL RIGHTS*   *Parts of the island were filthy with the waste from the glassmakers.* — EDWARD CHARLES,

27

DAUGHTERS OF THE DOGE   *He walked outside, out into the hot afternoon, sat on the edge of the grimy canal, stared at the rubbish in the water, lowered his face into his hands, and began*

FLOOD EFFECTS: For wintertime tourists an *acqua alta* can be a fun experience worthy of a few photos (1), the only hardship being an increased crush of people shuffling along in the dry areas (2). For native Venetians, however, it is a trial as they struggle to go about their daily business such as grocery shopping (3) punctuated now and then by having to rescue small children who can be carried to safety (4, 5, 6), leaving adults, even those worthy of divine favours, stranded (7). Some shopkeepers set out dodgy wooden ramps in a lacklustre attempt to lure in customers (8, 9), whereas others simply give up and wait until the flood recedes (10). Damage can be substantial though perhaps not to sealed items which are left in place (11; warning here: never, ever drink directly from a bottle or can bought in Venice) but certainly so for perishables such as flowers (12). Shopkeepers are therefore set scrambling (13) to raise their belongings on shelves (14) or put them back on boats (15). Meanwhile canalside restaurant seating loses its appeal (16, 17, 18) as workers make sure wiring is safely high and dry (19) or simply splash about waiting for low tide as they contemplate their lost revenue (20).

to weep. – DAVID HEWSON, *LUCIFER'S SHADOW* *The water below the bridge swirled with a sickly pea-green sludge sluiced up in the backwash of the propellers. There was a stink of putrefaction.*

— DAVID ADAMS CLEVELAND, *WITH A GEMLIKE FLAME*   *He took a last drag of his cigarette, stubbed it out, and flicked it into the Grand Canal. He smiled, 'It'll be washed out by the tide.'*

— EDWARD SKLEOPOWICH, *LIQUID DESIRES*   *The way was narrow and littered with broken roof tiles and moldering refuse.* — BEVERLE GRAVES MYERS, *PAINTED VEIL*   *Bonsuan finished*

SUMMER FLOODS: Recently, what had been unthinkable only a decade ago, summertime flooding, is now a regular feature of Venice's most flood-prone and touristed location, St Mark's Square. Here during high tides, water moves up through and emerges from the drainage holes (1, 2, 3) and spreads out over the Square (4), closing the Basilica (5) and the overpriced cafes (6). Meanwhile, undaunted and uneducated tourists doff their flip-flops and "refresh" their shopping-weary feet by wading in the sewage-contaminated water (7), even encouraging their children to join them (8, 9), and thus raising ethical dilemmas for knowledgeable locals concerning intervention. But because most Venetians despise with vehemence all tourists, good Samaritans are few and far between.

his cigarette and tossed the butt overboard, like most Venetians utterly careless about what he threw into the water. — DONNA LEON, A SEA OF TROUBLES   I loved to lie thus, dreamily combing

my fingers through the lagoon but the state of sewage was such inside the city that my nose killed the impulse even before my restless fingers began to twitch. – LISA ST. AUBIN DE TERAN, *THE*

WAVES: And, as if the flooding itself were not enough, Venice's structures are also subjected to a near constant pummelling of waves produced from motorboat wakes. The open water Giudecca Canal is a churning vortex for much of the day (1) due to passage of cruise ships, water taxis (2), and hundreds of freight boats that use it as a way to circumvent the increasingly common traffic jams chocking the Grand Canal. Waves from both these two large canals radiate outward into the heart of Venice along its narrow canals, causing splashing and building damage far from where the boats traveled (3), a situation made all the worse by speeding water ambulances (4) and firefighting boats (5). Although the *vaporetto* water buses seem to obey the speed limits, private taxis and personal boats often rush about producing large wakes (6, 7, 8). These waves roll (9) or crash (10) into the sides of buildings causing damage, then rebound back into the canals (10) serving to actually pull apart the battered foundations to such as extent that steel plating and reparations are frequently needed (11: note, wake from racing boat that has just disapped out of the left side of photo).

PALACE 'Of course it's an open sewer all this water. Just take a sniff. Filthy sods, aren't they? Wouldn't do back at home I tell you.' – REGINALD HILL, *ANOTHER DEATH IN VENICE Smells*

rise around me as if in Satanic prayer. This is the incense of the devil: offal of the streets, banana peel, dog shit, filthy candy and gum wrappers, dead migratory birds kicked into the corners. —

BUILDING DAMAGE: Venice may be a city of occasional falling debris from decomposing angels (1) as one popular book has phrased it, but it is really a city experiencing the ubiquitous transformation of Ruskin's famous "stones of Venice" into what might now be called the "scabs of Venice". High salinity floodwaters are destroying Venice's building heritage. The salt creeps upwards and expands, flaking off pieces of protective pink-coloured stucco and exposing the flesh-red bricks underneath, the whole looking ever so much like a gaping, festering wound…which of course it is. The process begins slowly with the damage restricted to a metre or two above the water (2, 3, 4, 5, 6), spreading upwards to the first floor windows and doorways (7, 8), and then to the second floor (9). Eventually the entire side of the building may be one big scab (10, 11, 12). The open wounds are now exposed to further damage. Cracks develop (13: upper right corner), bricks turn into a salty white pulp (14) and begin to fall off, first in isolated locations (15) and then in large lateral rows (16) or vertical holes (17, 18) that often occur around windows (19, 20) or doorways (21). Most alarming are the gaping holes that develop near the building foundations (22, 23, 24). Band-aid treatments include wooden splints across faces (25, 26), lattices holding up arches (27), and crutches sometimes seemingly holding up the very buildings themselves (28). Eventually, the entire structure becomes unstable and the building needs to be covered in scaffolding in preparation for its repair (29). Within this crumbling city there is no street with a more appropriate name than that of the street of death (30: note, missing plaster and bricks).

ERICA JONG, *SERENISSMA The canals stink like rotting corpses. Life is unbearable.* – ROBERT GIRADI, *VAPORETTO 13   In winter the canals give off a miasma that fills these ancient palazzo*

with odours of disease and disuse that herbs and rose-water cannot hide. Thus it is with Venice. The theatre of all our ceremonies, our splendor and fastidious courtesies stinks at times like a cemetery.

— JIM WILLIAMS, *SCHERZO* *As the game progressed, I lost some of my intoxication from my previous games and the whiff of sour canal water irked me with its undertone of urine hanging in*

the air. – LISA ST. AUBIN DE TERAN, *THE PALACE*   *This cesspit of a city, where sewers run like open veins* – SARAH DUNANT, *IN THE COMPANY OF THE COURTESAN*   *The wind was*

22

23

26

27

30

fresh and cold, coming in from the sea which gave no idea of what a horribly polluted place it really was. — IAIN PEARS, *THE TITIAN CONSPIRACY*   *Apartment complexes, factories, smokestacks,*

bridges, the chemical pall of terra firma were dimly visible in the distance all around this shore of this part of the lagoon in the gray light. — HAROLD BODKEY, *PROFANE FRIENDSHIP*

Com'era, Dov'era Imbrogliare:
Protecting and
Rebuilding Venice

"We must remind the worshipers of the lagoon
com'era dov'era (as it was and where it was)
that it has been continually reorganised
throughout the centuries in order to adjust to the
requirements of Venice's life and prosperity."

A. Rinaldo *in* Musu, I. (Ed.) *Sustainable Venice:*
Suggestions for the Future (2001)

"Imbroglio: From the Italian [of course]
imbrogliare, *meaning an intricate and*
perplexing state of affairs; a complicated or
difficult situation; a confused heap."

www.Dictionary.com

As the workforce and industrial output of the mainland site of Porto Marghera continues to decline, Venice has recently begun to examine the idea of redeveloping many of the abandoned warehouses and other structures. Real estate developers have created plans to reuse these complexes for such diverse purposes as industrial or environmental research parks, entertainment centres to reduce overcrowding at venues on the Lido during the film festival, much needed low-income housing, and education buildings for the city's universities. Surrounding all would be a series of large parks and open spaces. As well, a set of economic indicators has been developed to assess the reuse benefits of derelict buildings in the historic centre through the lens of applying sustainability concepts to urban revitalisation.

Whereas several international organisations have made the shift from being concerned about merely beautifying to truly saving Venice, others have persisted in their role as social clubs of well-meaning but naïve benefactors whose

> "experts worry about the proper shadings and the exact nuances of composition of various masterpieces [while] a good deal of Venice that Venetians continue to live in is falling apart." (Davis and Marvin, 2004)

In contrast to this criticism, the group Venice in Peril is now involved in restoring working-class housing.

Venice has realised that it could benefit from many more such projects. Toward this end, preliminary non-market economic analyses and health risk evaluations (based on soil contaminant mapping and surveys of cleanup technologies) have been undertaken with respect to remediating the 3,500-hectare brownfield site at Porto Marghera, which is one of the largest in Europe. Tentative plans exist for transforming another brownfield site in the neighbouring City of Fusina into a passenger terminal and new gateway to the lagoon. Concerns associated with redeveloping such derelict sites include the frequent lack of economic resources for redevelopment, as well as the significant issue of liability for contamination arising from past industrial activities. Without public support, private developers will not assume responsibility and need incentives to undertake the cleanup and reuse of marginal land of generally poor market value. Unfortunately, investment from Venice to explore such

regenerative projects has dried up in recent years as monies have been diverted to support MOSE (see below). Total reclamation costs for the entire area are estimated to exceed one and half billion euros. Not surprisingly, a comprehensive and integrated post-industrial regeneration of the entire industrial area of Porto Marghera is, at present, stillborn.

FORTIFYING THE LAGOON

THE first attempt to broaden understanding about the complexity of the problems faced by Venice occurred with publication of a special issue of *Architectural Review*. Published five years after the serious flooding of 1966, the magazine was significant in its highlighting of the need for the ideological descendants of Ruskin to move beyond their artistic preoccupations with salvaging individual buildings toward considering larger scientific and sociopolitical issues. UNESCO was one of the first organisations to recognise this, stating that

> "The preservation of a single work of art in Venice increasingly implies the repair and consolidation of the many elements of its environment, just as the man-made city of Venice itself can only be meaningfully restored in the context of the preservation of her much larger, more complicated and more sensitive lagoon". (UNESCO, 1979)

Later, Venice in Peril, realising that there was little point in repairing individual buildings when the whole city remained threatened, reinvented itself from an organisation focused on restoring monuments to one dealing with restoring the entire lagoon. Today, most observers recognise that Venice is very much a "diffuse city" with the lagoon being neither frame nor backdrop but rather an integral part of a single entity. For many, the only way to ensure Venice's future is to address the problem at the level of the lagoon. And, despite the tired chant of *com'era dov'era* by the building restorationists, with reference to the lagoon most scientists believe that the past is gone and that the present is non-reversible. The only option to maintain the health of the lagoon is therefore intervention. In other words, nature, if it is to persist, needs our help.

---

than a metre wide, up from which rose two steps, further evidence of the Venetian's eternal confidence that they could outwit the tides that gnawed away perpetually at the foundations of

Just Another Brick in the Wall

SEA WALLS along the Adriatic edge of the barrier islands of Pellestrina and the Lido were first constructed in medieval times. Since then, the sea walls have been periodically reinforced, including most notably a 20-km section of 4-m high and 14-m wide Istrian stone built in the 18th century. Embedded into this seawall was a plaque stating that:

> "The Guardians of the Water have set this colossal made of solid marble against the sea so that the sacred estuaries of the city and the Seat of Liberty may be eternally preserved".

It was the occupying Austrians in the 19th century who can probably take the lion's share of the credit for putting in place measures to protect Venice. They spent the equivalent of about $2 million each year on sea wall maintenance, including adding 15,000 tons of reinforcing marble per annum. Remarkably, until the implementation of the *insulae* and flood barrier projects in recent years, this Austrian intervention was the last major public works project to be undertaken to protect Venice from flooding. In contrast to the Herculean efforts by the Austrians, the Italian Government in Rome, with a history of generally ignoring Venice, has spent a meagre one dollar per metre per year to maintain Venice's defences. Not surprisingly, today's sea walls are in need of repair. In better shape are the other protective measures that have been implemented over the last decade, such as the reconstruction of 45 km of eroded beaches and the reinforcement of 11 km of breakwaters.

Today, some believe that the only long-term solution to Venice's flooding problems is to extend the sea walls across the mouths of the channels between the outer islands, thereby permanently sealing off the lagoon from the Adriatic. Construction of such permanent dykes would transform the lagoon into a freshwater lake and wetland but would of course also assure that the city would never flood again. While acknowledging the draconian aspect of this strategy with respect to lagoon ecology, supporters steadfastly argue that it is only by implementing such extreme corrective measures that any hope exists for solving the extreme problems of Venice. This is also the view expressed in the concluding remarks of the PBS Nova documentary which states that

"The Venice you know today cannot be preserved as it is today."

Others, however, have proposed an alternative protective measure that, though certainly more extreme in cost, might not be nearly so much so in terms of its environmental consequences compared to the permanent dyke approach.

The Big Biblical Techno-Fix

TWICE daily, four million cubic metres of water with an average tidal height of 70 cm flow into the Venice lagoon. Most experts agree that, given concerns about sea level rise, Venice's only hope for the future lies in its ability to seal itself off from the Adriatic when the need arises. Plans for building a system of protective floodgates go back to the 17th century. Today, about the only thing generally agreed upon with respect to building a barrier against flooding is that the challenge successfully to accomplish this will be large. Ideally, the requirements include the ability to deploy rapidly and reliably, the need for little maintenance, no interference to navigation and commercial port activities, an absence of, or limited effects on, lagoon water quality and ecology, and finally, when not in use, a structure that will be unobtrusive and have no impacts on site aesthetics.

Stymied for decades about how to satisfy these aforementioned concerns, it was not until the 1970s that engineers came up with the idea of building a system of mobile floodgates. What followed were years of plans, programs, politics, and procrastinations. A consortium of large Italian engineering and construction companies, the *Consorzio Venezia Nuova*, was created by a special law in 1984 to undertake construction of the flood barriers. A full-sized module of a single floodgate was built and hauled along the coastline in 1988 to gain public support, with the assembled crowd clapping as the mock gate was raised in place and attached to a large derrick. Referred to at the time as the *Modulo Sperimentale Elettromeccanico*, the acronym "MOSE" is now ascribed to the entire project, the name being a playful reference to Moses or Mose in Italian. The analogy seemed apt given the biblical prophet's ability to save his people through the act of dividing the waters.

Unfortunately, as a presage for much of what would later ensue, controversy began almost immediately. A public relations fiasco developed with the confused populace erroneously

the city. – DONNA LEON, *FRIENDS IN HIGH PLACES*   *Time eroded arches and quays, dark green water lapping softly at worn stone and crumbling red brick…From the inconspicuous*

47

believing that the large red towers used to suspend the single test gate in the demonstration project would be replicated for the final system of barriers which, as they had been informed, would consist of many floodgates.

Actually, the final design is comprised of 78 gates conjoined together at their base in three major groups which are to be situated across the mouths of the channel inlets between the outer barrier islands (Lido, Malamocco, Chioggia). Each bank of floodgates will be accompanied by a system of breakwaters and sea walls orientated in such a way as to reduce water inflows at each channel mouth by as much as fifteen to twenty-eight per cent. It is the mobility of the floodgates that, above all, is their most novel feature and perceived key for success. The design allows for the gates to swing back and forth with the waves, thereby dissipating energy to their supporting foundations which consequently need not be as large in order to anchor the array. Each floodgate will be about 20 m wide, 2 m thick, and 18 to 30 m tall, designed to hold back tides of about 2 m in height.

The ingenuity of MOSE lies in its intended modus operandi. When not in use, floodgates will lie prone on the bottom of the channels, nestled into their rigid steel bases. Given warning of an expected *acqua alta*, compressed air will be forced into the hollow gates, displacing the water that until that time has filled and weighted them down. The newly buoyant gates will rise up over a period of several minutes to block the incoming tidal surge. After the flooding threat abates, water will be allowed to flow back into the gates, which will displace the compressed air and cause the gates to sink back down to their beds. Here they will rest until called into duty again to block the next high water event.

The floodgates are expected to have a lifespan of sixty to a hundred years. Close to half a billion euros has been spent for the first phase of MOSE, with the estimated total cost to be in the neighbourhood of three billion euros. Construction is expected to take eight years and provide over one thousand direct and four thousand indirect jobs during that time. Up to one hundred and fifty specialised maintenance jobs will be needed thereafter.

Modelling exercises based on analysing flooding events between 1955 and 2002 estimated that the average closure duration of the MOSE floodgates would have been five hours for each *acqua alta* over this reference period. In other words, this would have been considerably less than one per cent of the total time, or about two to three closings per year, of which one or two would have been false alarms. Given a projected sea level rise of 20 cm due to climate warming over the next half a century, the average number of closings per year might rise to twenty-five, of which ten would be estimated to be false alarms.

The effectiveness of the MOSE system is influenced by a number of factors, including the amount of direct rainfall on the lagoon, the magnitude of stream inflows from the watershed, wind patterns in the lagoon and their effect on seiches, and also the structural operation of the gates themselves. To investigate the latter, a 10 million euro physical model of the lagoon was built in a Padua warehouse to test how the floodgates would perform under varying hydrodynamic conditions. This research turned out to be useful in detecting that, when waves of a particular pattern hit the barrier, adjacent gates will oscillate in opposing directions, thereby creating gaps into which floodwaters can penetrate. This discovery led to an alteration in the gate dimensions and angle of repose to counter the problem.

An international group comprising a *collegio* of experts (CELI) was appointed by the national government to review their own internally produced environmental impact study. CELI concluded that the MOSE proposal was "an effective way of protecting the city against high water" and the best option would be to proceed with the project together with continuing the ongoing *insulae* projects scattered throughout the city centre. Significantly, CELI also reviewed alternative flood mitigation measures which had been suggested by those opposed to MOSE. Interventions such as opening up all the fish farms in the lagoon, reducing inlet channel sizes, redesigning breakwaters, or creating/restoring mudflats, sandbanks and saltmarshes were deemed to provide a limited influence on water levels in the lagoon. CELI also concluded that other alternatives, such as raising up low-lying areas of the city to a height of 1.2 m (which is above the level of projected *acqua alta* incursions), would be just as costly to implement as MOSE, would take too long to construct, and in the long term would not be as effective as the mobile flood barriers. CELI also considered that any perceived operational difficulties with MOSE resulting from out-of-phase

---

tubular steel scaffolding around some parts it was obviously undergoing extensive restoration. – BEN HEALEY, *MIDNIGHT FERRY TO VENICE   Both of them had lived in Venice for*

resonance and gate wobbling, a sudden collapse of the entire barrier system, the difficult closure of the last floodgate against the increased pressure from the concentrated inflowing water, or any measurable leakage between the gates, would be minor and not issues meritorious of concern.

CELI concluded that the predicted frequency of about twelve closures each year, totalling only forty-two hours, mainly during the winter, would have negligible effects on the environmental quality of the lagoon. Only responses in relation to scenarios of severe climate change, with predicted gate closures of up to seventy times per year by 2050, were hypothesised to be capable of bringing about noticeable alterations in the chemistry and ecology of the lagoon. On the other hand – and one gets the impression that CELI might possibly have been grasping a wee bit here – it was thought that the floodgates might actually serve to retard the inevitable repercussions of climate change, their operational flexibility providing an allowance for the lagoon ecosystem to adapt gradually to rising sea levels. In this scenario, the floodgates would operate like taps or shock absorbers whose opening and closing could even be used beneficially to induce a circulation pattern in the lagoon that would flush out contaminants.

Finally, and most importantly yet most provocatively, CELI concluded that in relation to economic effects, the floodgates would provide measurable benefits that were equal to or exceeding their projected operational costs. However, with further sea level rise, the experts did acknowledge that the effects of frequent gate closure on shipping in terms of lost revenue to port activities would need to be re-examined.

Procrastination about implementing MOSE finally ended in the winter of 2000 when, after a series of severe storms, another notable *acqua alta* occurred which submerged ninety-three per cent of the city, the worst flooding to occur since 1966. Empirical modelling using data from this event suggested that the MOSE gates, had they then been installed and operational, would have limited floods to about half a metre above normal. Significantly, under these conditions, the floodgates would been closed for a mere nine hours and once reopened natural tidal flushing would have quickly and effectively cleaned out all pollution that might have built up during this short period of time. And so, with

support of a pro-business government, construction of MOSE finally commenced in 2003, more than three decades after it had first been proposed.

The Floating City

A methodology employed by the petroleum extraction and wastewater management industries might offer promise to help keep Venice above the tides. The deepwell injection of water or waste is often used to float hydrocarbons upward as well as to dispose of contaminants downward. This has led some to posit that a similar injection of seawater into the brackish aquifer located beneath the lagoon might actually be capable of raising the City of Venice by about 25 cm over a decade. Sceptics counter that the differential elevation responses which would be likely to ensue due to the presence of heterogeneous substratum could destabilise buildings. Groundwater injection was tried on a small lagoon island in the 1970s with disappointing results. Use of modern, more sophisticated techniques including a series of 600- to 800-m deep wells in a 5 km radius around Venice would, so the supporters believe, produce a uniform rise of about 30 cm relative to mean sea level. Other engineers remain dismissive that such a strategy would ever be effective on the scale needed to keep Venice high and dry.

Back to Nature

Saltmarsh restoration is believed by many to be a key toward building future resilience for the entire lagoon ecosystem. The process of restoring the saltmarshes of the Venice lagoon involves special challenges. First, as a result of a limited tidal cycle, there is a narrow window of suitable elevation in which to rebuild marshes. Second, reduced sediment inputs due to historic river diversions have caused the ground to become compacted. This in turn has led to permanently submerged areas which are in serious need of maintenance in terms of active sediment nourishment. Because redirecting the Brenta and Piave rivers back into the lagoon is unfeasible due to their present state of contamination, sediment is now actually sprayed onto the marshes from boats filled with dredge. Six million cubic metres of sediments dredged from canals have been used in this way over the last two decades to reconstruct nearly 1,000 hectares of saltmarshes and tidal

most of their lives, so they had an endless repertoire of stories about bribes paid to building inspectors or walls made of plasterboard that were pulled down the day after the inspectors left.

49

mudflats. And finally, as a result of increased motorboat traffic, it has been necessary to armour the borders of both the existing and newly restored marshes. Gabions, wood pilings, and bio-engineered fences constructed of wood bundles are used to entrap sediments for stabilisation and as breakwaters to reduce erosion from the propeller wash and boat wakes.

Today, reconstructed marshes comprise about fifteen per cent of the total marsh area in the lagoon. That is the good news. The bad news is that this is an amount that is roughly comparable to that which has been lost due to wave erosion and relative sea level rise over the same period. Reconstructing saltmarshes is therefore an ongoing activity just to balance out their rate of loss.

A series of regenerative landscape design projects have been completed or are in the planning stages. The ambitious reforestation project Bosco de Mestre has planted 1,000 hectares. Recreational trails and sustainable water management systems are incorporated into the reforested areas to improve the health, both human and environmental, of this mainland urban centre. Another Mestre project, Parco San Giuliano, is located on the site of an enormous reclaimed landfill. Here one can now ride about on a small train, play sports, attend outdoor concerts, sail, sit in cafes, and enjoy the birdlife in the stormwater treatment wetlands – all while taking in the wonderful views across the lagoon to Venice. Not only does this park provide the largest green space in the metropolitan City of Venice, it also offers an inspirational lesson of international significance in how to re-imagine and transform derelict land.

In the same spirit, but with an unlikely chance of ever being built, the *2G International Architecture Review* recently ran a competition to rehabilitate Sacca San Mattia, a nondescript, 31-hectare island immediately north of Murano. The goal of the competition was to re-imagine the derelict piece of land, presently functioning as a dumping ground for glass from the nearby furnaces, as a productive, regenerated landscape that would also serve as gateway to an hypothesised new "Venice Lagoon Park". The nature of such competitions is to generate interesting, out-of-the-box, ideas that will raise public awareness about issues rather than to produce immediately viable development recommendations. The danger in such competitions is that the posited designs are cheeky and clever rather than cautious and cerebral, and are simply too outlandish and ridiculous to capture citizen support. In such cases the public marginalises the proposals as being little more than weird art worthy only of being exhibited in the increasingly bizarre Biennale. Such a criticism is meritorious for this particular competition where, in addition to realistic ideas for windfarms, solar panel arrays, algae biofuel factories, community gardens, wildlife refuges, and tourist facilities, we have interesting but unrealistic ideas such as the island being completely covered in a canopy of tubes filled with lagoon algae that would strip away nutrients and pump oxygen back into the waters, a network of hiking trails throughout the lagoon upon which one uses float shoes to traverse across the water, a rail causeway to the airport, a system of warning buoys designed to change their colour in response to water quality much like colour-coded terrorist alerts, the partition of the lagoon into thousands of aquaculture cells, and all kinds of artificially created floating islands for sports fields, entertainment stadia, hotels, and eco-dome museum exhibitions.

Treating the Toxic Brew

IMAGES in the 1980s of gondoliers wearing face masks while paddling about Venice's reeking canals brought much adverse international publicity and finally forced city officials to admit that there was a pollution problem in the City. As a result, thousands of cubic metres of decomposing filamentous algal slime were harvested and removed from Venice's waterways. Since then, the lagoon-wide improvement in water quality has been one of the few success stories concerning Venice's environment that is worthy of celebration. Also, enormous headway has been made in development of contaminant transport models and the use of various rapid assessment techniques and biological monitoring protocols to measure and ensure the ongoing health of the lagoon.

The proliferation of algae in the canals was related to the eutrophication and elevated productivity of the lagoon. Improvements in water quality were due to reductions in nutrient inputs resulting from a ban in the use of phosphates within the lagoon. (This ban, it should be noted, only occurred in Italy decades after similar bans had been put in place elsewhere in the world. Indeed, Italy has a long tradition in playing environmental catch-up, as for example the fact that remediating post-industrial brownfield sites was only addressed in 1997, fifteen years later than in other western countries.) Further, the widespread construction of thousands of new septic tanks throughout the historic centre of Venice has reduced the amount of raw sewage being dumped daily into the canals. And finally, many fish farms have been reopened to the lagoon in order to improve flushing and consequent water quality.

Porto Marghera's industrial waste, thought to be responsible for the deaths of several factory workers in the 1980s, has for decades been dumped directly into the lagoon. Recently, in addition to curtailing new discharges, previously produced waste stored within badly constructed and leaking dumps has been contained within resealed landfills. In addition, three hundred thousand cubic metres of polluted sediment from 30 km of canals running through the industrial port, which had repeatedly been contaminating the water column through wind and boat traffic resuspension, have been removed by dredging. And to prevent the disastrous possibility of an oil spill, tanker traffic was finally rerouted away from the historic city and inner lagoon. Serious issues remain to be addressed, however. The obvious one is the continual discharge of untreated sewage from toilets and greywater from sinks and baths into the canals of the historic city.

Promising directions in pollution mitigation are beginning to be examined and implemented in Venice, decades after their application elsewhere. Some of these approaches use constructed wetlands, nature's kidneys, to remove nutrients *before* they enter the lagoon. The Ca' Di Mezzo complex of wetlands near Chioggia, constructed for water treatment, are among the largest of such systems that have been built in Europe. A major channel of the Brenta River that drains an area of extensive agricultural development higher up in the watershed is redirected to the treatment wetlands. Such "start at the source" technologies have a proven track record of effectiveness and are almost always less expensive to implement than traditional, "end-of-the-pipe" procedures.

In a similar vein, Venice is adopting green infrastructure such as small, artificial wetlands used to treat the waste from a community of fifty individuals on the island of Lazzaretto Nuovo. This wetland system, instigated as a pilot project that will hopefully inspire other lagoon communities, was built by a small consortium of grassroots organisations and is composed of individual septic tanks, beds of remediation or contaminant-sucking plants, a rainwater collection network, systems for wastewater collection and reuse, and a final polishing pond, which together serve to significantly reduce the outflow of nutrients to the lagoon.

Some of the harshest criticisms of the MOSE project come from those who fear the possibility of a buildup of contaminants when the floodgates are closed and Venice's centuries old "dilution is the solution to pollution" tidal flushing is prevented. According to the 2000 master plan for the region, an obvious solution to this problem is simply to collect and treat the sewage from as much of the population as possible. Toward this end, a new wastewater treatment plant is being constructed in the mainland city of Fusina. A complex of artificial wetlands is also being designed here for further "polishing" treatment of four thousand cubic metres of municipal effluent in addition to rain runoff. Interestingly, neighbouring industries will reuse this water as a substitute for their present withdrawal of water from the Sile River which, because of its higher quality, can be better used elsewhere. Additionally, this 100-hectare treatment wetland (one of the largest of its kind in Europe), unlike the one on the small Lazzaretto Nuovo island where space constraints were an issue, will include expansive open-water zones and sculpted islands that will attract and support a diversity of wetland fauna. Also, the treatment wetland will be surrounded by a landscaped park which will provide public education about wetland ecosystems and natural water treatment systems in addition to creating an area of recreational open space for the community. Despite a planning process that developed with the same glacier-like rapidity characteristic of most Venetian decision-making, this project signifies an important new direction in how Venice will address its environmental problems. Further, the fact that the project converts a post-industrial dredge spoil basin on the edge of the lagoon into an ecologically functional wetland with many ancillary beneficial uses for the wider community makes this the most important regenerative landscape design project currently underway in the Venice lagoon.

Another new development worth celebrating is the

MISS GARNET'S ANGEL   *'You'd think there was enough to do here…without asking for thousands of pounds to clean some statues!'* – REGINALD HILL, *ANOTHER DEATH IN*

51

construction in Porto Marghera of a power plant, the first of its kind in Italy and only the third in Europe, that will be run on a vegetable oil biofuel produced from algae. Forty megawatts of electricity will be generated, of which twelve will fill the energy needs of the port, with the excess being supplied to ships in the harbour. The Port Authority is quick to state that the technology will produce

> "an emissions-free energy source [which] would help preserve the historic lagoon city's delicate ecological balance."

Happy with the good public relations created by the project, the Port Authority is also considering building a photovoltaic park that would produce an additional 32 megawatts of solar energy.

VENICE You give cocktail parties and pat yourselves on the back because you have restored another painting and another church and the polite, cultured world of art lovers applauds you.

"LONG LIVE THE QUEEN"

SURVEYING: In order to know the precise degree to which the pavement needs to be raised to limit flooding and save the "Queen of the Adriatic" it is first necessary to undertake a comprehensive survey of elevations throughout the affected regions of the city (1, 2, 3).

Colour versions of all uncropped images can be accessed at **www.libripublishing.co.uk/veniceland**.

But what is the good of all your well-meaning efforts if Venice is going to sink in a lagoon poisoned by chemicals and oil waste or else be destroyed by a new flood? – RAYMOND RUDORFF,

WALKWAYS: With ample forecasting of an ensuing *acqua alta*, city workers scramble to erect an elaborate network of raised pedestrian walkways. Location priority is given to areas frequented by tourists for sightseeing (1) or shopping (2, 3) rather than residential areas occupied by locals. Occasionally water levels rise to a level to cover the much more modest walkways often placed in those areas off the beaten tourist path (4).

THE VENICE PLOT   All this land reclamation, all these ugly buildings, all this dredging. – WILLIAM RIVIERE, *BY THE GRAND CANAL   The last time he had been to Burano had been*

55

INSULA: Insula, the public works engineers in charge of repairing and raising Venice's pavements to protect the city from tidal flooding, has a number of projects underway (1). For those situated beside the open lagoon (2), steel barriers sometimes need to be erected and water pumped out (3, 4) before work can begin on the *fondamente* (5). One of the most visible of these projects recently occurred in front of the Ducal Palace (6) where storage of equipment (7), new underground piping (8), pavement stones (9), and actual work areas (10) were a necessary inconvenience to pedestrians. Insula is at work in many other areas of the city lifting up (11) and removing (12) the old pavement, laying down temporary walkways (13), replacing ageing infrastructure (14, 15), and stockpiling old and new paving stones (16, 17, 18).

in summer several years before, when its notoriously filthy canals were being drained. The odor had been terrible. — EDWARD SKLEPOWICH, *DEADLY TO THE SIGHT*   *A little past San Ivo*

a canal was being dredged, a dirty job saved for winter, when no visitors were here to see. Wooden planks dammed each end so big rubber hoses could pump out the water, leaving a floor of mud,

just a few feet down, where workmen in boots were shoveling muck and debris into carts. The mud covered everything, spattering the workers' overalls, hanging in clots on the canal walls, just below

SALTMARSHES: Emergency "sausages" of sand-packed fibre are used to buttress the edges of saltmarshes against further erosion from motorboat waves (1) until such time as more substantial arrays of wooden stakes can be set in place (2, 3, 4). These resemble miniature fortifications in some frontier setting that are erected against hostile natives, in the present case, these being the operators of the speeding motorboats. Due to the diversion of inflowing, sediment-laden rivers from entering the lagoon, the ecosystem is now starved of bottom material needed for the natural replenishment of eroded saltmarshes. Restoration practices therefore involve the presence of large machinery (5, 6, 7, 8) for the direct application of sediment that is either excavated and dumped or sucked up and sprayed upon the marsh. This built up material serves as the foundation (9) for recolonisation by native grasses (10). However, the continual bombardment of waves or swells from boat wakes against the wooden fortifications (11: note the rolling waves) eventually breaks them apart (12, 13) letting the invading water through the breaches to destroy the marsh once again (14).

the line of moss. — JOSEPH KANON, ALIBI   At the moment the canal was a muddy gulf in which a couple of men in rubber boots were shoveling out a channel. 'This ugly view is of course only

temporary,' explained the Signorina hastily. 'This sort of dredging goes on all the time, as you know, all over the city.' Stopping in front of a doorway, she produced a key and added learnedly, 'To

LANDSCAPE REGENERATION: The reclamation of an enormous landfill on the mainland in Mestre into the Parco San Giuliano, the largest park in the Venice lagoon, is a wonderful success story in the post-industrial reuse of derelict landscapes. Here locals can attend outdoor concerts, have picnics, engage in a variety of recreational activities (1, 2, 3, 4, 5), and observe wildlife in the wetlands constructed for stormwater treatment (6), all within sight of the historic centre of Venice.

keep the water moving as a preventive measure against high water.' – JANE LANGTON, *THE THIEF OF VENICE In the dank canal, the dredging has begun. Half a dozen men are up to their*

CANALS: Floating wooden planks (1, 2) denote canal restoration that often accompanies drainage. This work involves closing bridges and sealing off both ends of the canal (3, 4), moving pumps and other equipment into place (5), and buttressing any unstable building foundations (6) during draining of the section (7) containing decades of accumulated sediments (8, 9) that have inhibited the natural cycle of tidal cleansing and which in consequence will be removed.

wastes in the middle of the silt, black as demons, shoveling up the evil clods of sludge with cloths tied around their mouths to save them from the stench that the digging unleashes. – SARAH DUNANT,

62

IN THE COMPANY OF THE COURTESAN   The canal was sealed from end to end. – L. P. HARTLEY, SIMONETTA PERKINS   *Eventually, I end up in some foul part of town where my*

BUILDINGS: Repairs to buildings and courtyards, a ubiquitous feature throughout Venice, require the presence of barges laden with construction material that block canals for months at a time (1). Renovation of the foundations for buildings located beside large canals involves the use of steel barriers to create dry-dock conditions (2, 3, 4, 5, 6) and are often associated with Insula projects of raising pavement for flood protection. For repairs to other buildings, the small canal needs to be blocked and drained of water (7), thereby enabling workers access to reinforce the water-ravaged foundations (8, 9, 10, 11).

nose is assaulted by the stink coming from a drained canal, now a quicksand of mud. – SARAH DUNANT, *IN THE COMPANY OF THE COURTESAN   Forget about Venice sinking into the*

sea.– EDWARD SKLEOPOWICH, *LIQUID DESIRES* 'If only,' said the man from the Consorzio Venezia Nuova, 'we had had had been permitted to install our floodgates at the three points of

SHORELINE DEFENCES: Millions of kilograms of boulders are now used to protect the shorelines of islands throughout the lagoon (1), against motorboat waves, some of them being placed discreetly behind wooden cribs (2, 3) but most left in the open to become the new visual 'amenity' of the location (4, 5). Out on the barrier island of Pellestrina, rock and concrete jetties thrust out into the Adriatic in attempt to stabilise the shoreline against erosion and longshore currents (6, 7) at the same time as providing convenient places from which to swim for those locals and few knowledgeable visitors who refuse to venture to the filthy, crass beach at the Lido. The ultimate coastal defence system for Venice and its lagoon remains the massive bulwarks of reinforced stone, metres high and kilometres long, that armour the barrier islands (8, 9).

entry of the Adriatic into the lagoon, this problem would never again arise. As you are all well aware, our scale model has been working perfectly. We are ready to go.' It was a sore point. – JANE

LANGTON, *THE THIEF OF VENICE* *They [the Consorzio Venezia Nuova people] glowered at the mayor and one of them spoke his mind. 'It is the City Council which has prevented the*

LAGOON MODELS: Located in enormous buildings on the outskirts of nearby Padua are a group of hydrodynamic physical models that were instrumental in determining both how the Venice lagoon functions and the operation of the proposed MOSE floodgates. The lagoon model is a spectacular, to-scale creation that shows all the islands and channels (1) and, most importantly, can be filled with water (2, 3) to study the mechanisms of the inception (4), spreading (5), and flood peak (6) of different *acqua alta* events. Other, first-generation models were built to study how the MOSE gates would operate from closure (7) through deployment (8) against simulated incoming tides. Further, detailed studies conducted in a massive wave-propagation tank (9) were useful for observing the oscillating behavior of the deployed gates (10, 11) in relation to being struck by oblique waves, and thereby became instrumental for recommending important design modifications. Finally, a group of Harvard-Ca'Foscari university students standing on yet another, this time outdoor, model at the testing site joke around by simulating their own barrier (12) against the flood of propaganda that they have been subjected to from various spokespeople for the MOSE project.

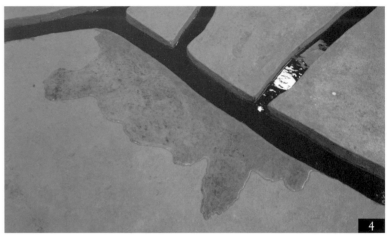

construction of the floodgates at the ports of Lido, Malamocco, and Choggia. Our hydraulic model at Voltabarozzo has been proven to work. When can we expect the permission and the funding to

begin in earnest?' 'Mister Fox has always been very generous with the city, I understand. Reconstruction. A lot of foundations are giving money away.' – THOMAS STERLING, *MURDER IN*

MOSE FLOOD BARRIERS: Massive building works for the accompanying jetties, breakwaters, and artificial islands that are underway in all three openings of the lagoon to the Adriatic, such as at the Lido (1, 2, 3, 4, 5, 6, 7), are necessary to redirect inflowing tidal water to the protective MOSE barriers whose construction has taken over the eastern end of Venice (8).

VENICE   'Do you really think all this money would be spent and all this concerted effort made if it were just a question of saving lives? Of course not. It is the objects that we really care about.

The floods of 1966, which were what really started all these international restoration projects and all these millions of dollars flowing in from abroad, they weren't such terribly serious floods as

DOOR BARRIERS: Another *acqua alta* band-aid is the placement of metal barriers across the mouths of doorways to prevent ground floor flooding, especially in those buildings which owners leave unoccupied for much of the year. Depending on the location and confidence of the owner, these barriers range from ankle (1) to mid-shin (2) to knee (3, 4) height and are even present on some of the major buildings along the Grand Canal such as the Ca' Pesaro museum (5) and the notorious Palazzo Dario (6).

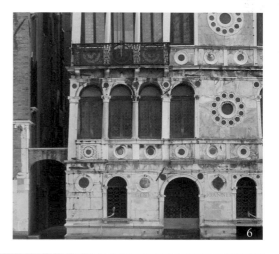

these things go. I mean, in the scale of cataclysms they would rank near the bottom. If it hadn't been Venice, if there had only been people to worry about, what would anyone have done? If it had

WASTEWATER TREATMENT: Construction of a series of treatment wetlands on the island of Lazzaretto Nuovo provides a new direction for how smaller lagoon communities can deal with their sewage. Nutrients present in the wastewater from restrooms (1) are removed by plants in treatment beds (2) before being discharged into the surrounding wetlands (3) for further polishing prior to entering the lagoon. It is somewhat ironic that this positive adoption of such innovative ecological engineering techniques based on living systems has occurred on an island so associated with death given that it was once a quarantine for the plague-ravaged city in the sixteenth century (4: garb worn by plague doctors now familiar as a carnival costume). It is also ironic that these remediation techniques that depend on contaminant-sucking plants should occur on the same island made internationally famous for the recent discovery of exhumed remains that had been imagined by the buriers to belong to a life-sucking vampire.

been ten times as bad, and somewhere else, what would anyone have done? I'll tell you. They would have sent a few food parcels and forgotten about it.' – BARRY UNSWORTH, *STONE VIRGIN*

73

Part Two

SOCIOLOGY

Chapter 3

Sempre Diritto, Seque Un Po La Corrente: Veniceland, Invasion of the City Snatchers

INTRODUCTION

SINCE the fall of the Republic over two hundred years ago, Venice has been a city defined not so much by its residents as by its pleasure-seeking visitors. Venice, once Europe's brothel, has now become "Veniceland", the world's favourite playground. It is the ultimate "eye candy" city, home to that unforgettable view of St Mark's Square with the Basilica, the Campanile, the Ducal Palace, the Molo lagoon edge, etc. being unequalled by any other amusement park anywhere. Tourism has become practically the sole point of reference for contemporary Venetian life. Indeed, R. Davis and G. Martin, authors of the important book *Venice, The Tourist Maze: A Cultural Critique of the World's Most Touristed City*, believe Venice to be in fact the world's first truly post-modern city, a Paris Hilton sort of place selling no product other than itself, where people flock for no other reason than simply to be part of the spectacle of

> "tourists on bridges taking pictures of other tourists in gondolas who are in turn taking pictures of them".

Venice, many locals have concluded,

> "has completed its long and sad descent from Queen of the Adriatic down to Belle of the Beach."

It has become a city that Debray called a

> "place of subsidised life, totally absent from itself, exist[ing] only in and for the gaze of others." (Debray, 1999)

SCHLOCK AND AH INVASION

VENICE has an undeniable Disney-like quality to it, "portrayed not as a present living city but as a phantom from the past". There are in fact two Venices, one being that of postcard fantasy, the other being that of overcrowding, decay, and discomfort. None of this is new. Historians point out that "practically everyone who has ever toured Venice has complained about the presence of tourists." For example, Henry James, the 19th-century misanthropic American writer and one-time Venetian resident, called tourists "trooping barbarians." And more than a century ago, Venice was referred to in derogatory terms as "pandering to tourists in her putrescence like an old whore". (Pertot, 2004)

Venice certainly is sinking, possibly not into the waters once MOSE becomes operational, but definitely into a rising and engulfing sea of mass tourism which many observers now believe poses the most serious threat to its survival. Davis and Marvin (2004) repeatedly use an aqueous metaphor to describe the situation: "a flood of foreigners" or "a human tidal wave" that displays a "rhythmic ebb and flow, pouring into the city" at particular times such that "when a tsunami hits town, the central areas of Venice are completely swamped" with St Mark's operating like "a tourist tidal basin".

"Tsunami" is an accurate reference given that a single cruise ship can dump upward of three thousand people (the equivalent of fifty tour buses) into the city and on some days up to half a dozen of these floating hotels may arrive into port. The recently published *The Venice Report: Demography, Tourism, Financing and Change of Use of Buildings* also uses the same aqueous imagery when they try to determine what the "saturation point" is for Venice with respect to accommodating tourists.

Venice is the world's most touristed city. Nearby Marco Polo airport has become the third busiest in Italy and is growing. The passenger port is now the thirteenth largest in the world and will soon be second in size in the Mediterranean. Estimates suggest that as many as eighteen million tourists a year now visit the city, three-quarters of whom are day-trippers whose numbers can exceed fifty thousand per day or one million a month during the peak season. Day-tripping results from the fact that hotel beds in nearby Padua, for example, are about a third of the cost of those in Venice itself and do not have the bad services characteristic of the old Venetian hotels that leaves luxury-seeking tourists disgruntled. The ratio of tourists to locals on individual days is now about 1:1 and annually is above 200:1. These are tourism levels that are nine times higher than those for Florence, for example, and which are unequalled anywhere else, including even Las Vegas, the world's second most visited city.

Most of these visitors are those pejoratively referred to as *turisti e fuggi* or "bite and run" tourists. The level of ignorance of these tourists – one cannot really even call them sightseers since for the most part the only sights they are interested in are those

She waited while a band of tourists in thrall to a furled umbrella trailed past. – JANE TURNER RYLANDS, *VENETIAN STORIES*   'Don't know how you can stand it. Is it always so

of each other posing – is staggering beyond belief. Many who drive over the causeway from the mainland are shocked to be unable to drive right up to and park in St Mark's Square. The bulk of tourists, if they have heard of the Bridge of Sighs at all, imagine it to reflect the blissful sighs of star-crossed lovers rather than the fatalistic moans of the recently condemned for which it was actually named. One tour guide interviewed in the documentary *Venice: Tides of Change* is particularly dismissive of the cultural depth of these individuals, stating that many arrive in St Mark's Square not even knowing if it is in southern or northern Italy and with a pressing interest to visit the restrooms of the Ducal Palace which "is probably the highlight of their day." More tourists actually go up the Campanile tower to gaze down at other tourists in the Square than enter either the Basilica or the Palace to see, much less study, the art and historic treasures located therein. The average length of time that tourists stay in St Mark's Square, termed by Napoleon to be the "finest drawing room in Europe", is less than an hour. And after waiting in long lines to enter the Basilica, one of the most historic and beautiful ecclesiastical buildings in the world, the average visit is only between ten and fifteen minutes.

And those statistics are for the tourists who actually do make an effort, no matter how feeble, to visit the sights of Venice. Almost unbelievably, fully sixty per cent of all cruise ship passengers do not even bother to disembark at all. These egregiously shallow individuals are more than satisfied to be able to check off Venice being "done" in their notebooks based on their brief viewing of most of city from the upper decks of the cruise ships during their slow drive-by. But then, who can blame them? Perhaps these are the wiser tourists after all. Who really wants to admit to being dumb enough to succumb to canal robbery and pay 80–100 euros for a forty-minute gondola ride which goes toward supporting a senior gondolier's annual earnings of 40,000 euros for his meagre three to four hours of work a week? Far better to stay aboard one's comfortable cruise ship and avoid crowds, greedy gondoliers, and for the most part, mediocre food.

The number of cruise ships visiting Venice has more than doubled in only seven years, up from two hundred in 2000 to over five hundred in 2007. At more than a third of a kilometre in length, these floating hotels truly boggle the mind. And because they bring in ten per cent of all tourist revenue to Venice, the city has bent over backwards to accommodate their needs. The passenger port is about to be expanded in order to be able to berth simultaneously five mega-ships whereas now merely three such along with a "small" one can be squeezed in. Repeated protests by citizens and even the mayor against allowing these behemoths to pass through St Mark's Basin, as well as measurements made of the resulting oscillations of building foundations, have been ignored. The Port Authority insists that they never want to deprive the passengers of the thrill of sailing past and looking down on St Mark's.

THE TOLL OF TOURISM

TOURISM has become the most pressing of all environmental and social concerns facing Venice.

> "For all its size and population, Venice probably suffers from more physical problems than any other city"

conclude Davis and Marvin, who believe that most of these problems are a direct consequence of the rampant tourism industry that has taken over the city. For example, although tourists provide thirty-five per cent of all economic activities on the island, they are responsible for eighty-three per cent of the amount of waste that is generated there, the disposal of which is paid for by citizen taxes. The amount of solid waste left behind in the historic centre by the crowd of a quarter million revellers after the final weekend of Carnival in 2002 was enough to fill St Mark's Square to a depth of 15 m, sufficient to cover the replica bronze horses perched on the Basilica balcony!

The litany of other environmental assaults that can be laid at the feet of the tourism industry are many. The once relatively ignored island of Torcello has now become the fourth most visited destination, and has suffered from massive renovations in order to enable water taxis and tour boats to penetrate right up into the heart of the island so the tourists do not have to walk in the half kilometre from the old boat dock. Back in the historic city of Venice, only the larger hotels with more than a hundred beds are required to have septic tanks; the others are given a

jam-packed? Just look at all the people! I'm surprised the whole place doesn't just sink plumb out of sight!' – EDWARD SKLEOPOWICH, *LIQUID DESIRES*   *'Even out of season,'*

79

waiver allowing them to dump waste directly into the canals, which they sometimes do down the *rii* away from their guests' sensitive noses but right underneath those of neighbouring locals whose concerns are not deemed as important. Further, those in the tourism business purposely go out of their way to disguise the city's problems, lying to tourists that the odour they smell emanating from the canals is due to decomposing algae or silt rather than that of their own raw excrement. Dredging of the canals, essential for maintenance and tidal flushing, was halted for decades partially due to tourist complaints about the smells emanating during the process. Local newspapers aptly refer to the Grand Canal as a motorway. The bulk of the twenty-five thousand daily boat trips in the historic centre (resulting in a boat passing by every six seconds) are a consequence of ferrying around tourists and restocking hotels and tourist restaurants with supplies. The city has continually failed to enforce both size and operating speed restrictions on boats, thereby allowing them to carry more and more goods and people faster and faster in order to feed the tourism vampire. Even the decision to build the hugely expensive MOSE barriers is considered by many Venetians to be pandering to the demands of the tourism industry rather than their own concerns. Finally, Venice's festivals, even those purposely designed for its few remaining residents, have been transformed into tourist extravaganzas in which the city is subjected to a schlock and ah invasion of city snatchers arriving in trains that, in the words of one newspaper editor, "vomit" upward of ten thousand revellers an hour. The degree of cultural hijacking is profound, with city officials in thrall to the decisions of television stations about where, when, or how festivals should be presented for the best possible "photo opps".

The monoculture of tourism in Venice impedes innovation and economic diversification, displaces almost all non-revenue-making activities, preoccupies city planners and decision-making authorities, generates major environmental impacts, and is in the long-run completely and unequivocally unsustainable. The independent and once proud city has been subjugated and enslaved to tourism. Venice has denigrated into becoming a contemptible "tattered Disneyland" (Musu, 2001). Many now share the pessimism that "there is little or no future in Venice, save for tourism" (Lauritzen, 1986). Some knowledgeable observers have sadly reached the obvious conclusion that the city should be abandoned to its unavoidable fate as an open-air amusement park existing solely for the benefit of tourists.

Venice was the most densely populated city in medieval Europe. One of the most powerful scenes in Luchino Visconti's film *Death in Venice* portrays a different city. Here we see the protagonist, played by Dirk Bogarde, shuffling about an eerie *campo* at night, completely alone except for smouldering piles of disease-infected clothes and splashes of lime disinfectant. Tourists have fled the plague-ravaged city leaving behind the existential, Camus-like hero as a solitary voyeur engaging in a melancholic danse macabre of cultural necrophilia.

Today, Mann's *Death in Venice* has become Fay and Knightly's *Death of Venice*. Tourism has torn apart the very fabric of Venice through creating a social degradation of almost unequalled magnitude. The soul of the city is slowly dying. Families are rapidly losing their connections with the city to such a degree that one former resident bemoaned that "nowadays the only way I can live in Venice is in my dreams". The most melancholic place to visit in the entire lagoon is the island of Torcello whose onetime population of thousands long since relocated to what is now the historic centre of Venice. The ghost town of Torcello may be a portent of Venice's own fate given the massive exodus of residents that has been occurring there since the end of World War II.

Since 1950, Venice has lost almost two-thirds of its population. From a post-war total of one hundred and seventy thousand, only about sixty thousand residents remain today. Most of those fleeing are the young, leaving behind the oldest population in Europe with an average age of forty-eight, of which more than one-quarter are over sixty-four. As a result, deaths now outnumber births and the two universities have experienced substantial declines in enrolment. Over the last decade and a half, Venice has continued to haemorrhage eight hundred people a year, a rate that is predicted to increase. Should this trend continue unabated, authentic Venetians will be extinct by 2030. The saddest sight in modern Venice is the "doomsday clock" located in the shop-front window of a pharmacy − one of the few, struggling local businesses that has not been replaced by a mask shop for tourists − that slowly ticks down the ever diminishing population of native Venetians. One urban planner recently stated "We've

Eufemia was saying, 'you'd be surprised how many tourists arrive every day in Venice.' − MURIEL SPARK, *TERRITORIAL RIGHTS*   *The streets between the Rialto and the Church*

reached the point of collapse, the point where things could fall apart."

When the population finally dipped below sixty thousand in November 2009, Venetians staged a mock funeral to mark the death of their city. A flag-draped coffin was carried through the city and floated in a historic boat along the Grand Canal accompanied by a barge on which a black-caped pianist played a sombre requiem on a grand piano. The flotilla made its way to the city hall where a poem was read in the native Venetian dialect. Meanwhile, students from an American university collected saliva samples from resident Venetians to record their DNA before the latter slid into the night to join the Neanderthals and other extinct hominids. One resident remarked that the funeral was long overdue, stating that "They came too late. We're already dead." Against this pessimism, Da Mosto and coauthors, in *The Venice Report*, maintain that "the city is not 'dying'".

Fully one-quarter of the Venetians who continue to live in the city actually commute to the mainland for work, where they cross paths with the daily influx of service workers arriving to the island to cater to the tourists. Most tourists have no idea just how few of the people who are serving them can afford to live on the island on which they work. In fact, from ten to twenty thousand people come into historic Venice to work every day only to return to their mainland or *terra firma* apartments each evening. One-third of Venice's workforce therefore lives elsewhere. The major proportion of Venice's daytime population is now composed of commuters and tourists. It is these demographic changes and imbalances that threaten the long-term survival of Venice.

Venice was slow to modernise. Three decades ago, only half the houses had bathrooms and those that did had toilets that drained directly into the canals where the waste joined that from emptied bedpans. In today's Venice, renovation costs are exorbitant and strict laws continue to make modernisation extremely difficult. There is a great attraction, therefore, in simply giving up on the whole decrepit place and relocating to the modern and much cheaper housing located in Mestre. A recent newspaper article, for example, was titled "Rising cost of living emptying Venice," and recounted the story of a born-and-raised Venetian who described the "trauma" of being forced to move to the mainland due to property prices in historic Venice being in the range of a million euros for a tiny apartment of just over a hundred square metres. House prices in Venice have indeed doubled since 2000, with the largest increase, forty-five per cent, occurring between 2001 and 2002 coincident with conversion to the new euro currency (thereby supporting the contention of many that price gouging during this transition period, which was regulated against in the rest of Europe, was simply ignored throughout Italy). Today what can be referred to as a new Bridge of Sighs is the causeway connecting historic Venice to the mainland over which departing natives have been forced to cross to assume their internment in the prison of Mestre, living their new desiccated lives like fish out of water. One elderly resident of the historic centre remarked at the recent mock funeral that "For an old person to move to Mestre, that is death."

Venice is filled with hundreds of abandoned and dilapidated houses. Even luxurious palaces along the Grand Canal stand vacant, their millionaire owners only infrequent occupiers. Can one really blame them? After all, though it might be a nice place to visit every once in a while, who really would want to live in Disneyland anyway, particularly during the summer when the tourists descend? No wonder that the noted Italian architect Aldo Rossi called for the city to be abandoned and converted into an open-air museum of monuments. Fay and Knightly's summation of three decades ago is apt now more than ever:

"Venice is a city beset by misfortune, suffering from severe neglect, caught in a cycle that is destroying her. Because she has been neglected, few of her citizens want to live there. Because fewer Venetians want to live in Venice there is less urgency to remedy the neglect."

Tourism has stripped the city of its culture and needed facilities, indigenous life being completely suffocated as tourists expropriate ever more and more of the city. The recent closing of the hospital on the Lido so that it could be sold for development, represented for many "the ultimate symbol of tourism over local interests." Venice is the only major city in Italy, for example, whose central town square, so vital for creating that wonderful sense of public place characteristic of Mediterranean urbanity, has been completely relinquished to tourism. Returning

of the Apostoli were crowded, making it impossible to go at any pace other than the prevailing saunter. – BARRY UNSWORTH, *STONE VIRGIN*   *Another flood of excited tourists*

native P. Barbaro expresses the loss:

> "The bridge beckons me toward San Marco; but I don't go to San Marco, haven't been there once in all the time I've been back. It is truly estranged from us now, our ancient historical centre, ever more alien to Venetians with each passing year. It has been invaded and occupied, sold off to the foreign troops at a loss. A memory assails me, of the last time I happened to come upon it, a few years ago: a circle of hell. We've abandoned it to them, the tourists, surrendered it up to strangers…San Marco has been desecrated, defiled. Little by little, we've removed it from the geography of our souls." (Barbaro, 2001)

Davis and Marvin pejoratively identified the "Bermuda-Shorts Triangle" of Venice as being that area conscripted by the Rialto Bridge, St Mark's Square, and the Accademia Museum into which nine-tenths of all tourists become trapped. This is similar to what Debray had earlier, and possibly more accurately, given that most tourists never enter a single museum in the city, termed the "fatal triangle", in which the iconic La Salute church is the third point.

> "Venice is not a real city, with real city's inhabitants and constraints, but a backdrop and a stage for one's gaze, emotions, and passions"

he wrote. The result has been that native Venetians have been relegated to being extras on a movie set, the few houses in which they live a part of the elaborate backdrop. Amazing as it sounds, many tourists are unaware that real people actually live in some of the buildings. It is not unknown, for example, for tourists to be found wandering about the upper floors of homes or private offices searching out and questioning those whom they believe to be actors hired as interpreters in some sort of living museum! This fiction of actors and audience is now actually marketed by cruise ships, who advertise that their 14-storey high decks are the best place to enable paying guests to "peer down on the city" whose small three to four-storied houses resemble a designed stage set around which the Lilliputian actors shuffle. It is just all so, so cute. And if it's not an open-air museum or a stage set, travel companies now market the historic centre of Venice as the world's most beautiful shopping mall. How can one maintain any

semblance of a normal life when living under such conditions?

There has been an explosion of rental accommodation in Venice to accommodate the tourist tidal wave. Incredibly, the number of Venetian properties converted into tourist lodging has increased one thousand and eight hundred per cent in just seven years, with more than 700 B & B's and pensiones springing up within the last few years alone, including a thirty per cent increase in the number of beds in just a single year. Today, one-tenth of all residential buildings in Venice contain tourist accommodation. Venice, often described as a "city-cum-museum" is in fact well on its way to becoming, with the possible exception of Las Vegas, the world's first "city-cum-hotel". This trend is expected to continue, the only difference being that it will do so at an accelerated pace given recent changes in the regional tourism laws specifically put in place in order to enable almost any Venetian building to be a possible candidate for conversion into a hotel. It will soon be legally possible for established hotels to expand into satellite buildings more than two hundred metres distant from the main structures. And if that were not enough, B & Bs will be able legally to reduce the minimum ceiling height of rooms that can be converted into tourist accommodation. Pessimists fear that the next step will be allowances made for reducing the surface area of permissible rental spaces, thereby enabling even closets to be rented out. Soon, sardine-stacked sleeping beds modelled after those in Japanese airports will probably make an appearance to fill the ever increasing demand.

Somnambulant residents are finally waking up to their bleak future. One citizen activist group recently staged a mass protest in St Mark's Square in front of a banner reading "Venice is not a hotel!" and handing out "Touristopoly" cards that portray the city as a board game in which all the historic sites are up for sale. UNESCO pessimistically summed it all up by stating that

> "We have now reached the breaking point for the preservation of a proper network of public-private services. Venice is becoming a museum-city, and is no longer a residential one."

Calling it a museum-city, however, probably does the tourists great credit given that, as mentioned previously, most never set foot in the Ducal Palace or any of the other cultural attractions around the city and really have no interest in history.

meandered beside the garden, buying trinkets at the souvenir stands and taking pictures of each other against the noble spread of the lagoon. – JANE LANGTON, THE THIEF OF VENICE

Unbelievably, some tourists disembarking from cruise ships to visit Venice do not even know they are setting foot in Italy!

Every year Venice becomes less livable for Venetians. Even in non-touristed areas it is often easier to find somewhere to buy a paper maché carnival mask than a bottle of milk or a newspaper. J. Rylands includes a heartbreaking short story in her collection *Across the Bridge of Sighs: More Venetian Stories* that recounts the travails of an elderly Venetian woman forced to ride the bus to the mainland to go shopping for supplies. Many are forced to make this shopping pilgrimage as the city's retail sector has been nearly completely taken over by masks, glassware, tourist kitsch, and fast food. With many buildings now occupied for only two weeks a year by absentee owners, there is simply no sustainable market to support local shop owners. The result is that it is possible to spend a day searching in vain for a hardware store in which to buy a hammer and nails. Even if one could find such a place and make the purchase, the oppressive restoration laws would probably prevent you from being able to do any simple home repairs…unless of course you are willing to pay the customary bribes.

F. Da Mosto asked in bewildering dismay once during his BBC documentary: "Are we in Venice, or Las Vegas?" It is a good and difficult question to answer, for at times it is true that the actual Venice seems no more real than the replica one on the Strip in the Nevada desert. Fay and Knightly are correct in their assertion that

> "Venice has become a city without a role, colonised by tourists and by Milanese and Turin industrialists. They use Venetian labor in their Mestre factories but they ignore the place their workers come from."

And that was written three decades ago. The crisis in Venice today – and anyone who tries to deny that there is a crisis is delusional – is really as much about people as it is about the non-human environment.

> "Everything must begin by bringing the Venetians back"

wrote de Combray more than thirty years ago. To put it simply, saving Venice in the end means saving Venetians.

I could never live anywhere else than here. Even if we fail to save Venice, I will stay here and go down into the water with my city. – RAYMOND RUDORFF, *THE VENICE PLOT*

CLOSE TO THE MADDING CROWD

TERRESTRIAL ARRIVALS: Tens of thousands of tourists arrive daily into Venice from automobiles, tour buses, and the train, crossing from the mainland via the bridge originally built in the nineteenth century by the occupying Austrians. Apparently, strategically located in every dozen or so of the arched openings underneath the surface of the bridge (1) are a series of chambers where, if deemed necessary, dynamite could be inserted to bring down the entire structure should Venice ever be invaded. Today, many locals are only half-jokingly examining putting in place just such a defence in order to stem the invasion of tourists. The western end of the historic island of Venice has been taken over by parking facilities (2) to accommodate the hordes. Tour buses cram a massive parking lot where disgorged luggage is conveyer-belted down into overloaded boats (3) to be taken away to the increasing number of hotels.

Colour versions of all uncropped images can be accessed at **www.libripublishing.co.uk/veniceland**.

While the crowds have thinned in the darkness, the city is rough-edged with people too drunk to care where they are going or whom they trample over to get there. — SARAH DUNANT, *IN THE*

MARINE ARRIVALS: Increasingly, more and more tourists rely upon cruise ships as their means for invading Venice. Traffic jams develop on weekends as departing or arriving ships jockey for position (1). The size of these leviathans, some a staggering fourteen stories in height (2), totally overwhelm all buildings in the medieval-scaled city (3, 4, 5). Nowhere is this more obvious than along *calli* in Dorsoduro when these moving mountains actually eclipse the sun (6, 7; note thousands standing on the upper levels). Just as alarming is to be interrupted when sitting at a desk or dinner table by a blaring speaker or choking exhaust fumes (8) and to look up over the top of neighbouring buildings and see thousands of tourists gaping down at you as if you were an actor on the stage of your own home (9, 10). One actually fears that when these monstrosities are tethered to the *fondamenta* edge (11, 12) they might actually be capable of ripping asunder the very fabric of the city itself should they ever topple over and capsize. One such berthing is particularly incongruous in that ships here completely dwarf the small corner house on the Riva dei Sette Martiri (13: bottom left of photo) which is actually the home of the explorer Giovanni Caboto who, in 1497, became the first European to "discover" North America (after, that is, the Vikings and possibly also the European Clovis peoples, Basques, and Irish but certainly not, as historically challenged Americans oddly believe, Columbus). Here one can walk beside the giant ships (14) and witness them open their bowels to release their next excreta of unwelcome tourists (15) who, like lemmings, blindly follow their umbrella-holding or flag-waving tour guides into the city (16).

COMPANY OF THE COURTESAN   *Before the day had far advanced, all the avenues of the great square were again thronged.* – JAMES FENIMORE COOPER, *THE BRAVO*   *He wanted to*

87

avoid the crowds clogging the main calli, making it almost impossible to squeeze through. – EDWARD SKLEOPOWICH, *LIQUID DESIRES* Tourists, many of them elderly, in their best summer

suits and dresses, moved along the pavement in reptilian slow motion. – IAN MCEWAN, *THE COMFORT OF STRANGERS*   *He could faintly hear a singer and an accordionist away on the*

LOST: Locals enjoy the pyrrhic victory in watching struggling invaders try to find their way around the complicated layout of the city (1, 2, 3, 4) though the geographical inquiries raised are always the same since most of the tourists never break free of the entrapment of the "Bermuda shorts triangle" of St Mark's Square, the Rialto Bridge and the Accademia Bridge or Salute church. At least the tourists pictured in these photos are independent and not part of the mindless hordes (see last section) in the tour groups who sometimes don't even know what country Venice is in!

Grand Canal entertaining gondola-loads of tourists, regurgitating Neapolitan sentimentality on to Venetian air. – WILLIAM RIVIERE, *A VENETIAN THEORY OF HEAVEN   In the past, heat*

TOURISTA ACQUA: Groups of tourists flow in from cruise ships and water buses along the *fondamenta* shorelines (1) and surge up in clogging white-capped swells over the bridges (2, 3, 4, 5, 6), often pausing in the most inappropriate places to take photos while doing so, before getting caught in the back eddies when window shopping in the narrow *calli* (7) until they eventually pour out into the huge tourist tidal basin of St Mark's Square. Here they gaze at the sights (8), enjoy staged spectacles (9), and a "few" spend hours (10) and hours (11) and hours (12) and hours (13) waiting in lines to enter the Campanile or Basilica, while the majority simply mill aimlessly about in vast crowds (14, 15, 16) taking photos of others taking photos of others who are in turn taking photos of them in one gigantic postmodernist experience of mass tourism on steroids. Locals no longer venture into the sucking whirlpool of St Mark's which they have completely relinquished to the tourists. They refuse, however, to surrender any ground when it comes to riding on the *vaporetto* water buses. Nothing raises the ire of native Venetians more than swarms of ignorant tourists who block entrances to the boats (17,18) and once on board insist upon standing like squashed sardines near the exits (19, 20, 21), some accompanied by their enormous suitcases or backpacks, for purposes of sightseeing. Finally, one after another of Venice's indigenous festivals such as that of the Redentore (22, 23, 24, 25; note cameras held high) continue to appropriated by rowdy tourists at the expense of the locals.

like this had reduced the number of tourists; now it seemed to serve the same purpose as heat in a Petri dish: the alien life form multiplied under his very eyes. – DONNA LEON, *DOCTORED*

EVIDENCE *The crush of people. They were everywhere. The bowls of Venice has opened up and lit forth.* – DAVID THOMPSON, *THE MIRRORMAKER* *The plebeians were dancing across the*

Piazza, lapping at the café tables and then receding, like the tides of the mother sea. Waves of sound thundered like breakers as they roared. – ROBERT ELEGANT, *BIANCA   Turning his back*

on the Piazza, he made his slow way through the clogged arteries surrounding the historic heart of the city until he reached the quieter alleys and squares and could breathe a sigh of relief. –

EDWARD SKLEOPOWICH, *LIQUID DESIRES*   '*With all the tourists, there's no room for Venetians.*' – DONNA LEON, *BLOOD FROM A STONE*   *The crush of people in the markets of the*

95

PIGEONS: Venice is also infested by thousands of "flying rats" who pander for handouts from tourists (1, 2) and occasionally express their gratitude by staging large thank you's (3) before flying away to their homes created by picking apart the already crumbling stones of Venice (4).

area. He broke through a particularly thick pack of people and burst into a small campo. – DAVID THOMPSON, *THE MIRRORMAKER* *'Can you see anything interesting out there?'... Just*

GONDOLAS: Everyone's romantic dream is of a solitary gondola ride on the placid waters of the Grand Canal (1) or along some lonely, mysterious backwater (2) but few realise the rarity of accomplishing this in today's Venice. Much more common are scenes of gondolas squeezing past each other in congested canals (3, 4) or large flotillas that form before heading out onto the Grand Canal (5, 6), a wise, protective strategy since the lone gondola (7) can fall easy prey to the packs of predatory motorboats that seem to hunt there.

people,' I said. 'Lots of people.' – CARYL PHILIPS, *THE NATURE OF BLOOD*   *'Quite a lot of people about at present,' Raikes said. Latimer nodded. 'They are like flies,' he said. 'The temperature*

TOURIST JUNK: Modern Venice has reinvented itself as one vast outdoor shopping mall selling the most horrendous kitsch imaginable. Tourists are lured by thousands of shops and kiosks (1 2, 3) that proudly display their "authentic" "Murano" glass or "Venetian" masks (most of which actually originates from China). Nothing has come to symbolize the death of Venice for locals more than the plague of mask shops that have taken over the entire city in the last decade (4, 5, 6, 7, 8, 9) pushing out other goods and services stores that are needed for preserving indigenous livability. Most of the eighteen million tourists who visit Venice each year are of the "bite and run" variety, there for only a few hours before they return back to the mainland or their cruise ships, leaving behind, however, mountains of waste to be cleaned up on land by city workers (10) or on water by environmental NGOs (11).

goes up a bit and they start to swarm.' – BARRY UNSWORTH, *STONE VIRGIN   Traffic was all there was at Piazzale Roma: cars, campers, taxis, and, especially during the summer, endless*

rows of buses packed there just long enough to disgorge their heavy cargoes of tourists. – DONNA LEON, *DEATH IN A STRANGE COUNTRY*   Colin made a vague, apologetic gesture, but the

BOAT TRAFFIC: From the top of one of Venice's campaniles it is often possible to count dozens of motorboats zooming about every which way (1). Although the water buses (2) are numerous, it is the near constant stream of boats that feed the tourism industry by carrying supplies for hotels and restaurants (3, 4) or other private craft (5) that cause most of the environmental damage and traffic congestion.

man, who was already walking away, enunciated with precision 'Tourists!' and waved his hand in special dispensation. – IAN MCEWAN, *THE COMFORT OF STRANGERS*

TORCELLO: With its abandoned canals (1), remnant buildings, and recolonising nature (2, 3), the lagoon island of Torcello, once populated by over twenty thousand people in the fourteenth century, may presage Venice's future if the exodus of the latter's citizens continues. And like Venice, extensive renovations are now taking place on Torcello (4) in order to make it more "tourist friendly".

The stallkeepers, the antiquarian and bric-a-brac sellers, the Murano glass trinkets, the fans and the watercolours were all filling the square, crawling over the carcass of Venice like hundreds of curious

EXODUS: As a result of financially ruinous rent, occasionally deplorable living conditions, and ubiquitous diminishing services, many boarded up homes exist in the historic center of Venice (1) as locals flee, not without palatable regret and genuine sadness, over the modern Bridge of Sighs (2, 3), abandoning their beloved city to the increasing floods of water and tourists.

maggots, leading parties of buyers into impossibly smelly alleyways there to tout their wares. — LISA ST. AUBIN DE TERAN, *THE PALACE*   *Venezia was full up.* — OSKIN OZCAN, *THE SECOND*

EXILE: Many thousands of Venetians have been forced to relocate to the sprawling bedroom community of Mestre on the mainland, a place of few distinguishing features (1, 2, 3) yet providing modern urban conveniences as well as other benefits such as the absence of flooding and manageable densities of tourists who remain contained in the area immediately around the train station. The sad reality is that it is actually possible to live a more authentic Italian life here in this concrete jungle than it is in the fairyland of Venice that has been sacrificed to tourism.

EXTIRPATION: Located near the Rialto Bridge is one of the saddest sights in all of Venice. An electronic tally board here is counting down the diminishing population of locals which had declined from 60,704 when in was established in late March 2008 (1) to 60,600 a mere three months later (2), a rate that if continued would mean that Venetians will be extinct within the next half century.

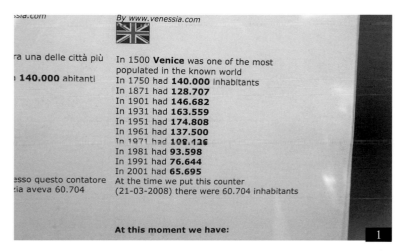

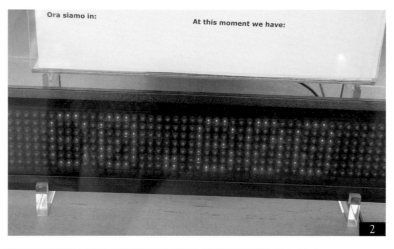

VENICE   Below in the square the tourists strolled around like pigeons and the pigeons ran like ants…Venice, he decided, was a film set. – REGINALD HALL, *ANOTHER DEATH IN VENICE*

"Rather than pity, this city inspires rage."

Brondi, C. *in* Pertot, G. *Venice: Extraordinary Maintenance: A History of the Restoration, Conservation, Destruction and Adulteration of the Fabric of the City from the Fall of the Republic to the Present* (2004)

"This disgrace of a city."

The director of a prominent NGO dealing with promoting sustainability in Venice, expressed in a private interview with the author.

"La Bella Confusione": The Beautiful Confusion

Once identified with a Fellini film, now used as a tenderly rendered colloquialism among those in the know to describe the resigned black humour needed to mitigate one's frustration at life in modern Italy.

Chapter 4

La Bella Confusione, Serenissima?: Petulance, Prejudice, and Politics in Saving a Dying City

INTRODUCTION

THE *Serenissima Republica*, or most Serene Republic, was how Venice fancied herself during her glory days. In recent years, however, the tone and temper of the debates about environmental and social issues have generally been anything but serene. Quite simply put, Venice excites huge emotions, and is an example of the very worst kind of petty power politics occurring within a fragmented, dysfunctional, and tension-filled governance system. None of this is new. The 17th century, for example, was filled with written polemics and soapbox speeches about the causes, consequences, and corrective measures that either should or should not be implemented to solve the problem of sedimentation in the lagoon. Debates today involve differing opinions on which structural and non-structural propositions might provide the best hope for keeping Venice above the rising tides of water and tourists, both of which threaten to engulf her.

THE TOLLS AND TOILS OF TIME

RESTORATION in Venice has always been a landmine issue. There are lessons for those engaging in environmental restoration today from the history of architectural restoration in Venice. Early debates raged, for instance, about the relative merits of conservation versus restoration, and how the latter may or may not differ from those other "r-words": reconstruction, refurbishment, and renovation. Never known to hold back criticism when he thought it justified, Ruskin admonished his contemporaries at every opportunity for what he perceived to be their glaring misconceptions and actions:

> "Do not let us talk then of restoration. The thing is a lie from beginning to end." (Ruskin, 2008)

He believed that the stones of Venice, which he considered to be the world's "most precious of inheritances", were deliberately being vandalised by "restorers" in their misplaced zeal to improve upon things. In point of fact, Ruskin considered that the actions of these well-meaning but ignorant individuals, carried out under the guise of restoration, actually ruined the very buildings they were endeavouring to preserve. Restoration, in this respect, might

be equated more accurately with mutilation. Perhaps the most glaring example of this was the controversial 19th-century repair work on St Marks's Basilica – described as "criminal" by William Morris – in which the names of the apostles were replaced with that of the restorer in several of the mosaics.

The restoration debate continued throughout Victorian times. The dismantling of the original Accademia Bridge, for example, became a metaphor for the battle between the old and the new. Then in 1910 futurists declared the independence of Venice from Italy based on their desire "to heal and purify this putrefying city" by filling in the "leprous canals" with the ruins of the palaces of the rich. Later, modernists entered into the debate about Venice's future with their own proposals for motorcars to penetrate into the very heart of the city along a drained Grand Canal Boulevard as shown in the scary special effects in the BBC show *Francesco's Venice*.

The concept of restoration continued to be regarded by the modernists throughout much of the 20th century as a way to rescue Venice's buildings from time's ravages by making the end-products appear to be somehow "better" than the originals. Others, however, championed the cause of celebrating time's relentless disintegration by preserving the picturesque decay for which the city had become so well known. Just as for today's ecological restorationists, the big question for architectural conservationists concerned the hoary issue about the importance of historical fidelity and which particular instance in time one should aim for in a restoration project. What was the ideal imagined pristine state? Was it before or after the Baroque accretions, etc?

Ruskin's opinions about the problematic nature of restoring buildings in Venice eventually became the accepted mindset. A 1984 governmental document warned of the dangers of "recreating instead of repairing", and a 1988 report was titled *Venezia "restaurata"*, the Italian word for restoration placed in inverted commas to indicate the futility of attempting to turn the clock back. Architectural re-creations, like waxwork imitations, ultimately have nothing to do with true restoration. Consequently, many churches today have actually had to be "rescued" from their heavy-handed earlier restorations through what might be called a process of "de-restoration".

Architectural restoration mistakes have and will continue to be made, however. The removal, for example, of the unsightly but

---

They say gates can be constructed to keep out the water which blows across the Adriatic. But I wonder if we can ever really turn back the tide. – SALLY VICKERS, *MISS GARNET'S*

106

protective black layer of accumulated grime from Istrian marble by well-meaning but naïve restorationists actually destabilised building foundations. And although use of steel and concrete may have enabled inexpensive repairs, unlike brick and wood, these materials have no elasticity to absorb the differential settling of the unstable Venetian subsoil and thus will likely contribute to future structural problems. Some architectural restorationists have recently shifted from using natural colours to harsher and garishly bright colours when painting repaired plaster. The result has been a transformation of the city, argue some critics, into an alien entity whose buildings are decked out like tarted-up beachside cottages rather than venerable structures. The recent restoration of the clock tower in St Mark's Square is a hallmark example of historic insensitivity in that the pieces were simply replaced rather than repaired and the latest family of caretakers, descendants of those who had resided in the tower for five centuries, were unceremoniously evicted. And then there is the irony involved with moving the famous gilded-bronze horses from the outside balcony on St Mark's to a confined corner inside the Basilica in order to "protect" them from the polluted air of Marghera. An irony given that the damage wrought by the excited breaths of thousands of gaping tourists each day might, if we are to believe some, be the more serious threat to these treasures of the Classical World. Cynics note that this decision had really less do with protecting rather than capitalising on the statues now that it is possible to charge admission to see them.

MOSE'S AMBIVALENT RELATIONSHIP WITH WATER AND PEOPLE

THE biblical Moses was defined by the variable nature of his relationship to water in terms of trusting his God. Today, it is the variable responses of an often untrusting populace that defines the nature of the modern MOSE project. Specifically, the central defining environmental debate about Venice's future concerns the question as to whether the MOSE project will either protect or damage the city and its lagoon. Citizens have become divided from their politicians over this issue. Matters are further complicated given that the majority of the metropolitan city's voters live on the mainland in Mestre, an area unsusceptible to

flooding and largely unvisited by tourists. Why then, these vocal voters argue, should they have to pay for MOSE when it does not affect them and is primarily being built for the benefit of tourists? Most of the information about MOSE is unpublished, difficult to find, and not in English, the international language of science. As a result, the opinions have been excluded of the community of those most knowledgeable and able to reach an informed conclusion about the effectiveness of the project. In consequence, little or nothing about MOSE has been peer-reviewed by a widespread group of independent, non-partisan scholars. Objective adjudication of the project has therefore been difficult. However, these limitations have not in any way hampered the generation of strong opinions by many about the efficacy and feasibility of building the mobile barriers.

Many of MOSE's proponents believe that the highly artificial nature of the lagoon necessitates major engineered interventions. They argue that, no matter how many water-absorbing wetlands are built, these will never be sufficient to mitigate Venice's flooding problems. Many supporters of the project are especially critical of polemical arguments "tainted by political agendas" that have curtailed a reasonable debate. There is no doubt that the din has drowned out more complex and subtle reflections about Venice's fate.

The cocksure and condescending manner with which MOSE officials state their case, as for example the chief engineer who is known for delivering information in a fashion resembling edicts (e.g. "without the MOSE, we will lose Venice"), galls critics no end. Historian P. Lauritzen accurately captured the opinion of many opponents of MOSE when he stated:

> "The arguments in favor of the barriers have been presented with such conviction, and backed with such an authoritative array of pseudo-scientific data, that this strategy has assumed the character of a foregone conclusion." (Lauritzen, 1986)

The Machiavellian way that the MOSE consortium has circumvented or marginalised the voicing of credible, alternative viewpoints has left opponents exasperated. The director of Italy's World Wildlife Fund goes one step further in his statement that the close links between government and big business in Italy have completely corrupted decision-making in Venice (for a

---

ANGEL 'Unfortunately,' growled the mayor, 'your hydraulic model does not take into account all the complexities of the situation.' It was a typical long-standing argument. – JANE

particularly telling example of this, see J. Berendt's interpretation of the soap-opera saga involved with restoring the Fenice opera house in his bestselling *City of Falling Angels*). Critics have accused the engineering consortium in charge of the MOSE project of accepting bribes from its member contactors. What big engineering firm participating in the decision-making process, these critics argue, would not want to build the floodgates and therefore do everything in its power to bring this about? Although the Venice municipality has legal responsibility for water management, this has been entrusted to an independent company. Even those whom one would hope to be independent of business interests, the governmental decision makers, are compromised by the way in which commerce and governance are interrelated in Berlusconi's Italy (as witness, for example, the alarming statistic that over ten per cent of all elected Italian politicians are actually *convicted* criminals) (Jones, 2009). As it is, much of the financial benefits accruing from constructing the flood barriers will go to the Consorzio's big business partners with head offices in Rome, Turin and Milan. (Venice in Peril. 2007–10)

Remarkably, an environmental impact assessment (EIA) of how the mobile floodgates would affect the city and lagoon was not ordered until 1995, long after planning for construction had begun. The result was that many "gate sitters", who until that time may have been ambivalent to MOSE, became hardened, vocal opponents. As J. Keahey stated in *Venice Against the Sea: A City Besieged*:

> "It is inconceivable that this step, common elsewhere in the developed Western world during the decade of the 1990s, took so long in Italy."

For many, it is this absence of an a priori EIA that is *the* inexcusable and glaring omission that severely compromises the validity of the entire project.

MOSE's opponents contend that the Consorzio's eleventh-hour, so-called "independent", review is a farce, given that of all people and from all places, it was undertaken by engineers from MIT. This, they believe, is like asking a fox if there is a perceived problem to the open door to the henhouse or a committee of elected Italian officials to adjudicate on whether there is corruption in Italian politics. "What else would one expect of MIT engineers?" asked one Venetian professor of urban planning. Even when it is pointed out that the review panel was in fact composed of accomplished individuals from several institutions in addition to MIT, critics retort that many of these were engineers, not ecologists, and that they were from the Netherlands, so again, what would one expect? The whole situation, opponents believe, is a teleological absurdity and foregone conclusion. Evidence to support such a contention, they argue, is right there for all to see. In 2000, for example, the regional administration authority issued a decree annulling a previous decree from 1998 that had been negative to the project. The fact that experts in the Ministries of the Environment and of Cultural Heritage had both been against the project, despite the blessing from the "independent" review panel, seems to have been conveniently ignored by the decision makers in charge, which of course rankles MOSE opponents like nothing else.

Environmentalists have been the most vocal opponents of MOSE, worrying that if the floodgates are closed too often the lagoon ecosystem will become irreparably damaged. The whole project is seen as needless tampering, engaging in a "disaster of doing." Engineers counter that everything one sees in the lagoon today is the result of massive interference by humans and not the product of natural evolution. MOSE, therefore, is simply part of a grand tradition of Venetian environmental management. Keahey, however, writing before the final decision had been made, cautions that

> "the recent history of Venice and its region is a history of failed engineering".

The creation of Porto Marghera in the 1920s within sight of the sublime historic centre of Venice was in his mind, "strike one", and the failure of a nearby dam in the mountains that resulted in the deaths of thousands in the 1960s was "strike two". MOSE, in this baseball analogy, if built, would be "strike 3" in terms of poorly conceived engineering projects. And the debate goes on.

Some mathematical models suggest that the gates would lower flooding tides by 20 to 30 cm whereas other models believe that MOSE will work for only rapid in-and-out tides and not those that linger for more than a full day. Several of these latter models even go as far as to suggest that, because the high river discharge and watershed runoff which contributed to the *acqua alta* in 1966 originated from the mainland and not the Adriatic, the serious flooding would still have occurred even had MOSE been operational at that time.

The World Wildlife Fund is opposed to both the mobile barriers as well as the accompanying permanent earthwork structures erected at either side of the channel mouths, which they believe will severely alter the flow of tides and currents, thereby "provoking consequences of unpredictable proportions". Proponents of MOSE intriguingly (and perhaps somewhat teasingly, one imagines) retort that the barriers could actually be exploited for the betterment of the lagoon's ecology. The strategic and staggered opening and closing of the three sets of floodgates in relation to tidal cycles could, so the engineers advance, be used to induce a circulation pattern to flush out pollution and thereby actually *improve* the water quality.

Given all these uncertainties, critics steadfastly regard MOSE as a bad decision, based on putting all the eggs in a single basket of unproven effectiveness. Not true, say MOSE's supporters, stating that the barriers are merely one element in a comprehensive plan involving a suite of interventions throughout the entire lagoon and city. The *insulae* projects, for example, are thought to be able to protect the city from water levels up to 1.1 m above sea level. At higher floods, MOSE would kick in for further protection. In this reasoning, Venice can be saved only through investments in *both* the barriers and in raising the city and restoring the saltmarshes. Posing one against the other is to set up a false dichotomy. In the end, many knowledgeable and objective observers concede that the mobile gates will most likely meet Venice's present-day problems but might not be adequate to address future concerns. Consorzio spokespeople admit as much, stating when pressed that the "barriers are never the final solution in a changing world but would buy a century of usefulness". (CELI, 1998)

The key question in deliberating the effectiveness of MOSE boils down to concerns about the projected frequency and duration of barrier closings in response to climate-induced changes in sea levels. Official sources predict that there will be about seven closures a year, an amount that most experts agree will not produce a buildup of contaminants in the lagoon. Although most closures are thought to be of a relatively short

duration, questions persist about the effects of repeated closures over the entire season. Some critics dismiss impact studies that have failed to consider the seasonality of the flooding threat, in response to which MOSE engineers argue that, due to the low biological productivity in the lagoon during the winter, floodgate closings at that time will have no influence on lagoon ecology. But remember that Venice is slowly sinking at the same time that the sea is persistently rising and the number of floodgate closings will certainly have to be increased dramatically to protect the city.

Understanding flooding in Venice is weakened by short-term perspectives both in analyses of the problem and in predictions about the proposed solutions. Because of this, A. Ammerman and C. McClennen regarded predictions made by Consorzio engineers to be sheer hubris. Extrapolating forward based on an examination of the historical rate of land subsidence determined from archeological evidence suggests an increase in relative sea level height of 30 cm over the next century. Significantly, this is a value that is seven times larger than the low scenario value used in the three official, and what many believe to be overly-optimistic, impact reports. Models based on what critics consider to be more realistic projections about sea level rise predict that the floodgates could be closed for as much as ten per cent of the total time, or for twenty per cent of the winter period. Obviously, the implications of such a substantial duration on the consequent buildup of contaminants cannot be dismissed. By 2050, according to some independent analyses, the floodgates might be deployed for much of the wet season, in contrast to the seven closures per year that MOSE proponents predict. In this scenario, the floodgates could be closed up to twenty-four times each winter month, in other words, about one hundred and fifty closings, many of which would go on day after day for four long months. No one knows what this might do to lagoon ecology.

At best, MOSE will allow Venice to buy some time in relation to the "inconvenient truth" of global climate change. The question remains, however, about how long it will be until the floodgates can no longer provide that protection. A series of simulations which, unlike some official scenarios, include leakage from wobbly gates resulting from being struck by obliquely oriented waves, hypothesise that MOSE will become obsolete within only a few decades. The barrier system will then have to

the elements. Totally forlorn quest. — BARRY UNSWORTH, *STONE VIRGIN*   *The Venetian capacity for dawdling is of the largest.* — HENRY JAMES, *THE ASPERN PAPERS*   'We talked

109

be demolished to enable the construction of a more effective separation of the lagoon from the sea. As a result, even among those who agree in principle about the predicted efficacy of MOSE, some would conclude that the whole project might very well be "a terrible waste of taxpayers' money."

Some environmental scholars would be much more amenable to accepting the MOSE solution had the project been seriously examined from the triple bottom-line of sustainable development which considers and gives equal weight to both social and economic factors in addition to environmental ones. As I pointed out in *Handbook of Regenerative Landscape Design*, with the exception of I. Musu's *Sustainable Venice: Suggestions for the Future*, which was published in 2001 before MOSE became a fait accompli, there have been few attempts made in this regard. An exception was an investigation undertaken by several groups of Harvard and Venice-Ca'Foscari university students. One group determined that ultimately it made better economic sense to take the money otherwise spent on building (two and a half billion euros) and maintaining (twelve million euros a year) MOSE and invest it in the stock market. Over the projected optimistic sixty-year lifespan of the floodgates, this investment could generate forty billion euros, an amount that is more than adequate to pay off the estimated four million euros necessary to compensate individuals and institutions for flooding damage each and every year. (It is important to note, however, that this economic analysis does not consider the difficult-to-quantify discomfort brought about through having to live with repeated flooding.) Another group of students concluded that the absence of all of an adequate competitive bidding process, a thorough and accountable cost/benefit analysis, modelling that gives accurate credence to the future predictions of climate change, and finally a fair-minded consideration of alternative solutions in a more comprehensive assessment, meant that, in their opinion, the project should be judged as a risky, unwise endeavour. "MOSE" in this case, the group concluded, could really double as an acronym for "Major Obstacle to a Sustainable Environment".

It would be wrong simply to dismiss these sustainability analyses as being generated by naïve students, as the supervising professors of each class are internationally regarded experts in the field. And they are not alone in their criticisms. When groups of international experts review MOSE, such as the international Peer Review for European Sustainable Development, the conclusions reached are markedly different from those generated by the Consorzio's handpicked reviewers:

> "The contradictions in this project are numerous, among those most great is the irreversibility of the action. The exorbitant costs compared to the limited benefits in terms of high water events (e.g., the barriers can be closed a few times a year for tides over 1.10m while St. Mark's Square floods at 80cm) have caused a part of the scientific and political world and that of environmental associations to be against this project, thus proposing less expensive, reversible, effective alternatives."

The fact that the MOSE project is now under construction has not assuaged the debate. Proponents maintain their steadfast conviction that there will be no perceived operational nor construction problems from the mobile gates to the lagoon environment. Blaming MOSE for anything to do with the quality of water in the lagoon is misdirected, they insist. Rather than being obstructionist reactionists, environmentalists should really do a much better job at proactively addressing the causes of water pollution by working toward more effectively managing or regulating the agricultural runoff from the watersheds, industrial discharges from Porto Marghera, and sewage from their very own Venetian toilets. These are the main culprits, and if one was really concerned about the environmental health of the lagoon, then it is the irresponsible farmers, factory managers, and city officials whom should be called to task, not MOSE's well-meaning engineers. There is of course a powerful argument here since, in the absence of any pollution in the lagoon, many (but not all) of the concerns about the hypothesised effects of temporarily sealing off the lagoon from the Adriatic would disappear.

DRACULA, BAAL, MOSES AND MEPHISTOPHELES

JUST when it seemed that questions about whether or not to build MOSE had finally been laid to rest with commencement of construction, the future of project could again be in question. Insula, the public, and of course, this being Italy, the private conglomerate responsible for implementing the *insulae* projects

about the rescue work in Venice, which apparently he considered was now too slow and too late.' — BEN HEALEY, *MIDNIGHT FERRY TO VENICE* 'They rape our land. Destroy our sea.

throughout the city, is supposed to be an equal partner with MOSE. The reality is quite different. Because most economic resources are funnelled to MOSE, there is little left, so the head of Insula complains, to repair the city's infrastructure (actually Venice's funding for maintenance has been reduced by one-quarter in recent years).

"The problems of sewers in the city are just as important as high tides,"

he states to deaf ears as the group considers legal actions that would halt MOSE by forcing an injunction against the present unequal distribution of funds. The mayor, supported by the majority of the city council who are opposed to the project, backed a proposal that would also temporarily stall the project until due examination is made of its disproportionate receipt of restoration funding as well as its blithe dismissal of inexpensive alternatives such as inflatable rubber "sausage" barriers.

And then there is the European Union who may also be entering the fray. Arguing that restoration funding should be spent on building maintenance restoration rather than solely on flood barriers, they have also been threatening to take Italy to the international court unless a proper environmental impact statement is carried out. So, whereas MOSE's ultimate need and effectiveness may be worthy topics for debate, it is impossible to deny that its development has created a giant and ravenous vampire that is sucking the lifeblood of funding out of the coffers needed to run the city. The financial health of the Venice that remains resembles that of the hapless victims in those Hammer films from the 1960s: a drained pseudo-corpse, neither fully dead, nor completely alive.

The search for alternative funding needed to keep the city functioning has had one serious consequence observable to both residents and visitors alike. As recently as half a decade ago, Venice was blissfully free of almost all advertising. Today, however, lack of funding has led the city into a Faustian deal with big advertising, "Big" here means both big revenues ensuing and big eyesores resulting. Most city buildings under restoration now sport large advertisements affixed to the outside scaffolding, sometimes for the duration of the entire project which, Venice being Venice, may drag on for more than half a decade. The advertisement that (as I write) covers the façade of the Correr Museum in the Piazetta San Marco, for example, is 240 square metres in size, half the area of an Olympic swimming pool. And bigger ones are on the way! In exchange for this visual spear thrust right into the historic heart of the city, Venice receives €3.5 million for the building's needed restoration from an advertising agency that rents out the space for €50 thousand a month (rising to €75 thousand a month during February's Carnival). Noted economist and Venice critic J. Kay does not buy the city's justification about the necessity of their Faustian deal with advertising companies. Venice's monetary woes, he states, are a result of chronic financial mismanagement by the city, the shortage of money being symptomatic rather than causative of the problem.

Shared condemnation of this overt consequence of the MOSE vampire is possibly the sole issue that has residents siding with tourists. There is actually a bylaw that states that advertising on restoration signs must not "detract from the appearance, decorum, or public enjoyment of the building." How ludicrous and hypocritical then to have over-officious police chastising and discouraging ignorant tourists from eating fast food in St Mark's Square because it is "disrespectful to the place", while those tourists sit on marble steps beneath an enormous advertisement that has been approved by politicians and which pictures some tennis star or James Bond villain wearing a Swiss watch.

Fatalists, however, shrug off criticism about the visual blight, saying that the historic core of Venice has long since been converted into an outdoor shopping mall, so advertising, no matter how large in size or seemingly incongruous, is really no big deal, especially if it is needed to keep the buildings standing upright. In short, Venice has already been sacrificed to the Baal of mass-market tourism and there is nothing left that is culturally worth protecting anyway. So what if millions of tourists sigh over their "ruined" photos of the advertisement-draped Bridge of Sighs? They should be happy that it is still there. And, if they don't like it, they can always stay home…*please*.

Meanwhile the mayor, in his continual and desperate search for money, is considering leasing space in *calli* and *campi* throughout the city which would allow Coca-Cola exclusive rights to install their vending machines. Where will it end? At what point does, or did, the Queen of the Adriatic become the whore of Babylon?

Poison the soil.' — STEVE BERRY, *THE VENETIAN BETRAYAL*   Venice hides its shame. — LAURO MARTINES, *LOREDANA: A VENETIAN TALE*   The special UNESCO representative

BOFFINS, BUFFOONS, AND BUMBLING BUREAUCRATS

IF a single phrase can be used to explain the inertia of governmental decision-making in Venice, it is "bureaucratic lethargy". Venice's real "Achilles' heel," noted one scholar "is not fire and it's not high water. It's bureaucracy!" Although a series of special laws have been passed over the last three decades to make protection of Venice a national priority, critics maintain that really little has been accomplished. As P. Piazzano summed it up:

> "No other city in the world has been studied in such detail. None has been so painstakingly dissected to determine the reasons for its rise and fall. And, it must be added, never has so much hard work produced such meager results." (Piazzano, 2000)

The reason, he believes, is the built-in flaw in Special Law 798, the most important piece of legislation enacted to protect Venice since the flood of 1966. The law states the requirement to "restore the hydrogeological equilibrium of the lagoon, slow down and reverse the process of degradation and eliminate its causes" at the same time as the need for "preserving the area's productive and economic interests." To Piazzano, it's as if

> "the law's authors seem to be the direct heirs to the [Venice] playwright Carlo Goldoni and his Harlequin who served two masters." (Piazzano, 2000)

One can envision an episode of *Yes, Minister* about the impossible absurdity of the whole thing.

There is also a perfect storm of other reasons to explain the lack of progress on solving Venice's problems: the inoperable and paralysing bureaucracy so characteristic of Italy, the production of grandiose yet pointless plans that require no action being undertaken, a fatalistic acceptance of corruption as a way of life in Italy, and an exodus of sensible Venetians in the face of invasions of insensitive tourists. Despite plans existing for four decades to stem emigration to the mainland by providing restored buildings at low rent as well as decent schools and hospitals, none of this has occurred in any widespread and effective manner. Such procrastination, a form of tortuous death by a thousand plans or the "production of plans by means of

plans", follows a long-established Venetian tradition. An emblematic example of this concerns the famous Accademia Bridge. In 1933, the original Austrian iron bridge was disassembled and replaced by a temporary wooden structure. After more than a half century of fruitless discussion and inability to make a final decision on the permanent design, this wooden structure had deteriorated to the point that it had to be torn down, wherein – no surprise – another such "temporary" structure was put up. In short, Venice is plagued by overly cautious policy-making that often leads to acceptance of the status quo of doing nothing rather than risking the possibly of making an incorrect decision.

Plans for restoring or saving the entire city have fared no better. There is a blackly humorous scene in the Nova documentary on the "sinking" of Venice: a woman is shown sitting at a desk with a telephone receiver in each hand and accompanied by the narrator's voice saying "Since 1966 Italians have been talking,,,,and talking…and talking…". A few minutes later, viewers are informed that the MOSE gates have been a political hot potato that has been tossed from one city administration to the next, which in Venice has meant an incredible thirty-five times. Citizens, the documentary suggests, are becoming increasingly angry, one disgruntled local referring to politicians with the statement "blah…blah…blah…." However, when one puts this into an historic context, the three decades of deliberations about MOSE are minor given that one and a half *centuries* of discussion were necessary before any action was taken with respect to diverting the sediment-rich Piave River away from the lagoon.

Following the 1966 *acqua alta*, Italians spent nearly four decades bickering about what to do to fix the problem. During this time, increasing numbers of foreigners grew more and more impatient and frustrated at the country's seeming inability to save its primary cultural jewel. As described in the McCurry documentary *Venice: Tides of Change*, northern Europeans and North Americans became apoplectic about the "disconcerting nearsightedness by [Italian] government administrators". Had the consequences not been so dire of the repeated failures by various Italian governments to address adequately the problems of Venice, it would have the making of a farce in which the incompetence

who had been discussing the Venice problem with the Italian government and their special committee, urging them to make good their repeated promises, returned to Paris in a furious

112

of the players is scarcely believable (again, think *Yes Minister* here). International and Italian experts alike agree that, as a result of the structure of governance in Venice and the inherent lack of transparency, accountability, and legitimacy of the different groups involved, plans for solving the city's problems almost never become operational in any satisfactory way.

The end result of all this is that many practices that would be unthinkable and unacceptable by the legal and ethical standards of any other Western country are accepted and commonplace in Italy on the whole and in Venice in particular. Not without reason is Italy near the bottom of the list for developed nations based on favourability rankings for doing business. In consequence, the ensuing foreign criticism has been insulting, almost vitriolic, and quite possibly, also meritorious. For example, from J Kay:

> "The problems of Venice are not pollution, technology or finance; they are problems of politics, of organization and of management."....."A sad series of accidents has placed so many of the best of Western Europe culture and civilization in the hands of Europe's most dysfunctional political system."

Fay and Knightly are no gentler:

> "The very nature of government in Italy, its inherent instability, its system of political favors, and its crushing bureaucracy, made it unsuitable to handle a problem like Venice, and it is the basic reason for its spectacular lack of success ......The fact is that politics in Italy *are* different, and politics in Venice...are unusual even by Italian standards.";

and

> "There is an illness in government today and Italy has it worse than most. When a politician takes power then he also must take responsibility. The pleasures of power are among the rewards for taking that responsibility. But in Italy the people who take power decline responsibility".

The presence of fragmented and overlapping institutional responsibilities between various administrations (and even among departments within a single administration) in Venice has led to a Balkanised and ineffective system of institutional governance. This, more than anything, is what observers identify as being the major impediment to solving Venice's environmental problems. Such a jumbled morass of multi-tiered governance is simply following the long-established Venetian tradition of making the complex, complicated, as for example the process for electing the *Doge*:

> "In a nutshell: 30 men were selected by lot from the *Maggior Consiglio*; they reduced themselves by lot to 9 members; they elected 40, who reduced themselves to 12, who elected 25, who reduced themselves to 9, who elected 45, who reduced themselves to 11, who elected 41, who finally elected the doge – 25 votes was the winning number. This rigmarole could last quite a while." (Buckley 2004; from Morris 1960)

No wonder the word "imbroglio" has its origin in Italy.

The "woeful Italian tradition" of governmental malfeasance has long been recognised. Many jokes exist about Italians changing their governments more frequently than the rest of the world changes their light-bulbs (it is actually 58 governments in 55 years). The sad state of affairs concerning the environmental mismanagement in Venice, however, is no laughing matter. In 1979, UNESCO, finally losing patience, put in writing what its members had long thought and railed about in private. Reference to the blunders and negligence that were a "characteristic" feature of Venetian and Italian governments fuelled the international fires, leading to calls for a foreign *doge* to be brought in to run the city (possibly as a private city-state such as the Vatican) in order to save it from Venetians. The suggestion by some international critics that the responsibility of governing Venice be given to UNESCO was of course an affront to Italian sovereignty. In retribution, the agency was temporarily run out of town like the bad guys in an old spaghetti-Western.

Venice, in addition to suffering from ineffective governance, is also plagued by the endemic corruption that lies at the core of modern Italy, as described by T. Jones in *The Dark Heart of Italy*, his realistic antipode to the bucolic *Under the Tuscan Sun*. Phrases that have been used to describe this characteristic feature of the Italian economy include "mysterious", "secretive", "suspiciously complex", and "defies forensic interpretation." In Venice, this corruption interferes and often undermines all restoration works. It is believed to be accepted practice that a complicated system of bribery needs to be followed before any building can be

temper. – RAYMOND RUDORFF, *THE VENICE PLOT* 'But more tourists, I suppose,' Brunetti said, referring to the deity currently worshipped by those who ran the city. – DONNA

113

modernised. In terms of the environmental restoration of Venice, one-time promised loans were never given, and of the 550 million euros that were initially targeted for MOSE, much was either moved to other projects or mysteriously disappeared into the secret pockets of an incredibly corrupt system.

This dismal history of financial mismanagement and graft has led many international observers to become harshly critical of Italians' ability to save Venice. Fay and Knightly, for example, summarise the entire sorry affair as being "a tragedy within a farce." The extent of greed of the Italian government seems to be in a class of its own compared to the rest of Europe, enabling them without any embarrassment to actually tax (at twelve per cent) all international and national donations given to restoration *charities* in Venice. For many this rapacious affront, which has been a serious impediment to restoration activities, became the last straw, leading to the suggestion that Venice needed to be saved not for, but *from*, the Venetians and Italians. Fair-minded Venetians are themselves just as critical about the ruinous state of affairs in their city and country. The founder of the Central Institute for Restoration, for example, raged at the blight caused by shortsighted industrial greed:

> "Marghera is neither a city, nor a community, but a jumble of factories, pipes and chimneys, facing Venice, which looks on appalled at the monster it has pupped. Therefore Venice is dying by its own hand, and too many of its citizens deserve the fate of Martin Falier." (Pertot, 2004)

Martin Falier, it should be noted, has the dubious distinction of being the only *Doge* who was executed.

It should be no surprise that Corruption Perception Index surveys measuring "abuse of power for private gain" have shown the Italian public to have, with the exception of Greeks and Bulgarians, the lowest expectations of quality from their public services and servants in all of Europe (Transparency International, 2010). The legendary financial mismanagement and skullduggery of Italian politicians and business leaders – not that there is much of a difference anymore between those two professions (Jones, 2009) – has led to repeated calls for the international governance of Venice. Recently, *The Venice Report* weighed in with their conclusion that

> "Though it is hard to foresee the protection of Venice being entrusted to some abstract global authority, we see a financial role for the EU, whether directly or through the European Investment Bank." (Da Mosto et al., 2009)

In the end, Venice's government often seems, at best, to be at cross-purposes with, or at worse, actively to act against, the wishes of its populace. It contributes, either by incompetence or greed, to what G. Pertot in his seminal survey *Venice: Extraordinary Maintenance* called "the widespread rape of the area". One way partially to counter this problem, he believed, is to engage more fully "the actors who, until now, have remained somewhat excluded from the political debate."

VISIBLE PROBLEMS, INVISIBLE PEOPLE, AND SUSTAINABLE RHETORIC

A true story:

> As the public became worried about increased contaminant levels resulting from creating an embayment sealed off from the sea, vocal opposition to the installation of the flood barriers continued to grow. Recognizing the need for more environmental impact data before proceeding ahead with the innovative project, construction was temporarily halted. The ensuing public consultation process was found to be extremely useful for improving the final effectiveness of the flood barriers. One official stated that: "Our overall conclusion is that the key factor in the success of the public information process has been the open character of the process, the systematic analyses and the free and full availability of the analyses whereby its results and conclusions have been available to all." By "all", this official meant local municipalities, city governments, NGOs, the scientific community, interested individuals, and the press. Further, what was referred to as the "open character of the process" involved making the environmental impact assessment documents available for commentary on the web, with hard copies placed in libraries, as well as production of summary brochures, news articles, and meeting report minutes…

The unfortunate truth is that this anecdote, though certainly true, refers to the story of the flood barriers constructed in St.

Petersburg, Russia, not those in Venice. Venice's process for reviewing and finally deciding upon its own flood barriers, rather than being open and egalitarian, as was the case in St. Petersburg, was instead secretive, technical and exclusionary. The irony of course is that these are essentially the same adjectives that have long been used pejoratively by the rest of the world to describe machinations within the old Soviet Union.

It was not always this way, however. At one time the governing body created a special commission of patricians to decide how best to strengthen Venice's embankments based on the advice obtained from laborious interviews from the local people, the *proborum hominum*. Today, if there is a single problem involved with determining the future of Venice, it concerns the need to engender a shared vision of the city as a social as well as a material system, something that can be brought about only by finding ways in which actively to engage people in the process.

The seminal book *Flooding and Environmental Challenges for Venice and its Lagoon: State of Knowledge* is based on a series of workshops held at the University of Cambridge on such topics as urban flooding, engineered solutions to storm surges, physical-chemical processes in the lagoon, hydrodynamic modeling, lagoon morphology, and water quality. The focus of the book was on scientific issues and, as a result, only a few contributors discussed sociological concerns. One author recognised this weakness, stating his belief that studies by scientists and engineers are inherently weakened by glossing over or ignoring the human elements in Venice's struggles against the sea. Another author considered that engineering solutions alone could never be expected to solve the long-lasting problems of the lagoon. Besides the scientific factors of biology, chemistry, and morphology, human elements such as culture, history and socio-economic concerns should be given equal weight in any decision making. A third group of authors most clearly stated the need for engaging the public:

> "And with regard to policy making, the 'polymorphism' of the city and its lagoon, given its long history of man/nature interaction, and the vital underlying natural dynamics that maintain the system, demand the wider participation of a well-informed public in decision making." (Fletcher and Spencer, 2005)

One inescapable truth that emerges from the decades of polarising debate about the MOSE barriers is that both the Venetian public and the concerned international community have needed to be kept much better informed about the multitude of factors affecting the health of lagoon and the preservation of the city. When, for another book, I reviewed effective watershed management projects from around the world, one de rigueur cornerstone of success they share is a high degree of ecological literacy among their respective citizens, something that was fostered by environmental communication. For Venice, a positive development in this regard was the production of the companion book generated from the Cambridge meetings: *The Science of Saving Venice*, published in both an English and Italian edition. Highly illustrated and with a text pitched at about a *National Geographic* level of detail, the book is designed to fill in the gap between debates amongst experts and the concerns and interests of the general public, as voiced in the press. The MOSE consortium is itself currently engaged in a catch-up program of technology transfer. At their drop-in information centre in central Venice, interested parties can obtain an attractively illustrated brochure and poster explaining the need for the mobile floodgates. In addition to this useful propaganda, there are information boards erected around the city in association with many ongoing *insulae* projects.

But is public education alone enough to help Venice survive? In order to be truly effective, watershed management must really be based on a public that is willing to move beyond being informed about an issue to becoming an active participant in socio-environmental reparation and healing.

In a summary chapter at the end of the book *Flooding and Environmental Challenges for Venice: State of Knowledge*, the authors consider the city's future in terms of the three pillars of sustainability. For the social pillar, they conclude that Venice must develop low cost or subsidised housing in order to restore a demographic balance to city. For the economic pillar, they believe that a system of tourist user fees needs to be implemented. And for the ecological pillar, they endorse the MOSE barriers. A surprising omission in the nearly 700-page long Cambridge book (in addition to the relative dearth of material about public engagement noted above) is that exclusive of this last of the sixty-

ALTA   *There could never be enough people helping Venice in its war against decay, pollution, and the hungry sea – not to speak of governmental apathy and bureaucratic ineptitude. –*

eight chapters, sustainability, according to the index at the end of the book, is mentioned only eight times. Fortunately, a book was published four years earlier on the subject, *Sustainable Venice: Suggestions for the Future*, which fills in some of the gaps.

Many observers believe that preserving essential features of a lagoon ecosystem is the key to formulating solutions concerning Venice's long-term sustainability. A social consensus on how to implement the various proposed solutions must involve all stakeholders with an interest and an opinion on the problem of Venice (locals, nationals and internationals). The City of Venice has finally come to the realisation, long held by its legion of international critics, that its future is uncertain and compromised due to the lack of a comprehensive, coordinated, and integrative development plan. Toward this end, they are investigating how to create conditions for local sustainable development as part of their governance system through which to ensure social, economic, and cultural quality for the community. The good news is that the plan is designed to identify potential scenarios for the city, to encourage cooperation among stakeholders, to prioritise interventions deemed necessary to facilitate the objectives, to promote responsible use of diminishing resources, and to suggest approaches for monitoring implementation of any projects that are proposed, all in the light of European standards. Although these laudable goals are being pursued in the Action Plan, an international review panel selected to access the performance of Venice, in terms of the existing problems and potential roadblocks to achieving success, identified a number of issues that needed to be resolved.

One serious impediment to the adoption of a sustainability vision for Venice resides in an ill-informed and possibly ill-formed city council. In particular, there is thought to be a leadership vacuum at middle management positions in city governance whose individuals are characterised by ignorance about sustainability issues in general and who display a lack of commitment in moving toward goals of the plan in particular. The review panel phrased it thus: "There is still organisational resistance and signs of organisational inertia in the field of sustainability" (PRESUD, 2004). Difficulties also exist in the division of responsibilities between the chamber of commerce and the city that do not encourage sustainable development, as

well as economic growth initiatives that are not based upon decoupling growth from production of resources and reduction of waste.

Other issues needing to be resolved include: performance management that is difficult to achieve when few targets and service standards have been set; no evidence of a comprehensive approach to analysing different options in policies and programs; a need to adapt integrated impact assessments that reflect the nature of sustainable development; investment in a new sewage collection system that remarkably does not include treatment of waste; lack of clarity around roles, responsibilities, and accountabilities of the four separate government agencies administering water management in the lagoon; a need to change the policy in which households are charged directly for waste services they receive but eighteen million tourists are not; the public perception that green space is left over rather than a vital resource in its own right; and the absence of a clear and strategic approach to integrating environmental issues in the economic redevelopment strategies for the post-industrial future of the port of Marghera. In the end, one is forced to conclude that Venice's politicians are involved more with sustainable rhetoric than they are with sustainable development.

In Venice, a palatable sense of community powerlessness exists due to lack of public engagement in the sustainable development agenda. This has prevented people from identifying and therefore owning the issue in such a way as to move toward changing their own behaviors. The review panel stated that

> "unless Venezia starts to enable stakeholder input through a bottom-up process then we fear the progress with the strategic plan will be hampered by lack of broader ownership and co-ordination". (PRESUD, 2004)

As a result, there is an absence of progress on stakeholder involvement and an absence of a transparent public commitment to involve the public in the inception and evaluation of ideas.

In Berlusconi's Italy, wherein the mayor of Venice has been reduced to being little more than a lobbyist to big business hegemony, "Who", asks the Chair of the Venice in Peril Fund in frustration, "can speak up for Venice?" The answer must and can only be its own citizens, as discussed at length in several chapters

RAYMOND RUDORFF, *THE VENICE PLOT* *'It's frustrating. I sometimes think that more publicity could get things moving. Unwelcome publicity, I mean. For example, press*

in *Handbook of Regenerative Landscape Design* including "Visible Cities: a Meditation on Civic Engagement for Urban Sustainablity" by R. Abbott, and "Reparative Paradigms: Sociological Lessons for Venice from Regenerative Landscape Design" by myself. In one section of this book, "Public Participation, the Key to a Sustainable Venice?", I examine four concepts for fostering public participation: the Local Agenda 21 sustainability protocol arising from 1992 Rio Summit; the Vision for Venice initiative promoted by the NGO Forum per la Laguna; derivation of alternative futures scenarios in comprehensive land-use planning; and the physical engagement of people in contributing to the restoration work. In summary, Venice, if it hopes to survive into the next century, needs to do a much better job of constructing methodological *passerelle* or pedestrian bridges that would allow its citizens to have a greater opportunity to walk across the rising waters of discontent and to be able to participate in and help to set a sustainable course for the years ahead.

WILD LIFE MANAGEMENT

TOURISM has become the self-created Frankenstein monster that Venice can no longer control. J. Brodsky, in *Watermark*, his loving homage to Venice, came down in favour of gates as being the only viable solution to Venice's problem with flooding. The Nobel Laureate poet differed from the scientists and engineers, however, in just where he would locate the protective barriers. Rather than out in the lagoon between the islands and against the Adriatic, Brodsky believed that the gates should be erected at the exit from the Santa Lucia train station in order to protect Venice from the flood of tourists. Others have voiced the same opinion. At a recent debate in London on whether Venice should be allowed to die, noted economist J. Kay posited that "the sea of tourists may be a lot more threatening than the Adriatic."

Unfortunately, relatively few attempts have been made to raise consciousness, to study, or to try to save the city from the rising tide of tourism. *Vaporetto* or water bus workers regularly stage wildcat strikes that shut down the city in protest against the burgeoning increase in traffic from service boats carrying supplies for the tourism industry that makes their jobs increasingly difficult. Gondoliers in turn regularly block the Grand Canal for a day or more in protest against all motorboat traffic (though one feels that perhaps in this case their grievances might have as much to do with concerns about lost foreign revenue than concerns about environmental problems associated with propeller wash). Nothing enrages locals more than *vaporetti* that are clogged with tourists and their enormous accompanying suitcases and backpacks. Newspaper editors have publicly confessed their fantasies and almost uncontrollable urges about hurling the offending culprits into the water (no editorials ever written have garnered the same amount of sympathetic support from local readers). Finally, in 2008, a new *vaporetto* service began that is restricted to holders of a Venice residency card. Ruefully, one wonders if this was a preemptive move by city bureaucrats, not so much to appease their own citizens as to forestall the looming and inevitable scenes of battling Venetians and visitors whose consequent negative publicity might have deleteriously threatened the inflow of tourist dollars. One positive step is a study (tellingly undertaken by a group of foreign university students rather than city officials) that identified the potential for greatly reducing the amount of boat traffic in the historic centre through the creation of goods transfer stations whereby loads among smaller boats could be consolidated onto a few, larger craft heading to the same destination. Another promising direction is the Venice Connected Program, begun in 2008, which allows tourists to pre-book admissions to cultural sights while discouraging them from visiting the city during the busiest periods. Then there is the newly established Istituzione Parco della Laguna, a city agency created in 2003 and tasked with enhancing the "urban void" of the northern lagoon for, at the same time as paradoxically protecting it from, tourists.

Due to an incompetent legislature, a weak regulatory system, and the absence of incisive measures, "public intervention to regulate this situation [of rampant tourism] does not appear to have gone further than declaration of intent". Realistically, the city has regulatory control of only the parking garage and bus services as means to limit inflow, with no long-term agreement in place with Trenitalia to limit the number of arriving trains that bring in most of the tourists. Only on a few holidays is the bridge to the mainland closed to stem the number of vehicles (including tour buses) from entering the city. But this protective measure is only implemented at the point when crowds of over one hundred thousand are expected. Many believe that by that time, it is already too late, the damage having been done.

Other attempts to limit the ravages wrought by tourism have either been circumvented or ignored. Hotels and restaurants, for example, hire firms to pump out and remove their sewage. Unfortunately, it is commonly believed, and sometimes supported by evidence, that unscrupulous firms engage in the illegal dumping of the waste in remote corners of the lagoon. There have been some successes, however. Removing the pigeons from St Mark's Square had always seemed unlikely given that "Venice wants to keep the tourists amused, and the tourists are amused by pigeons". Despite this, after years of legal wrangling, it is now illegal to sell pigeon food in the Square. Even here nothing is simple, since no sooner was this bylaw enacted then animal rights groups started to feed the "poor, starving birds." And so it goes on.

It is impossible to ignore the reality that Venice is drowning under an unmanaged flood of tourists. As such, the time has come to find ways to manage those tourist flows. The big, unresolved, gorilla in the closet is whether visitors to Venice should be charged some sort of head tax or entrance fee. Time and again the tourism industry, with support from the Port Authority, has successfully squashed attempts to impose a tax (even minor ones of €5 a person) on hotel guests and cruise ship passengers that would have supported preservation of monuments and city maintenance. The industry questions the fairness of their customers having to pay such a tax when the millions who pour freely in and out of the city through the train station are the ones responsible for most of the damage. Okay, the counter argument goes, perhaps the time has finally come then for a truly "radical approach", based on charging *all* tourists the actual price of consuming the natural resources and disposing of or recycling their waste when they visit Venice.

Almost half a century ago, the novelist M. McCarthy, in a series of highly influential articles and later a book, likened Venice to being "part museum, part amusement park, living off the entrance fees of tourists." A decade later, R. de Combray continued to invoke the theme-park image of Venice:

"You are tempted to settle, in Venice, to watch it become a museum…It was the palaces, the architectural wonders: they needed the attention, she claimed…As she spoke, I had an image of a fabulous Disneyland. You would enter a turnstile and find a city without inhabitants, spotted here and there with sumptuous palaces filled with great works of art. Possibly the guides would be costumed…" de Combray, 1975)

Today, as J. Martin correctly states in her interesting book about Venetophiles, *No Vulgar Hotel: The Desire and Pursuit of Venice*:

"It is impossible to have a conversation about the city without Disneyland being invoked the way preachers invoked Hell."

Because the economic returns to Venice from day tourists are minimal relative to their damaging role in contributing to the demise of the city, observers since the early 1980s have seriously suggested that admission be charged. Kay believes that Venice should be managed as a tightly regulated theme-park:

"Today 12 million people a year pay 50 euros to visit Eurodisney. It is clear…that if the Disney Corporation owned Venice, Venice would not be in peril."

Elsewhere, Kay clarified these provocative comments, enunciating that Venice would benefit from being run in the same caring fashion that Disney manages its properties, rather than actually being run by the corporation itself.

Carrying capacities exist for all tourist destinations beyond which the pressures consume site resources and irreparable damage ensues and visits begin to decline. Some life-cycle models of tourism development have suggested that Venice can withstand a critical limit of about twenty-five thousand tourists a day, of which fourteen thousand are day-trippers. The stark reality is that this upper threshold is now exceeded more than two hundred days a year. Venice may therefore soon reach the point where Yogi Berra's classic maxim begins to make sense: "No one goes there any more, it's too crowded." Would that this would be the future! Unfortunately, no matter how congested, how chaotic, how truly unbearable the situation becomes, the crowds will keep coming and coming. Venice is just too hard-wired into the world's collective psyche to be ignored. People will always visit Venice; for as one tourist wrote in *The Times*: "Venice is a congested,

the perils to which it is now exposed because of neglect, apathy, and the criminal conspiracy of political and financial interests. – RAYMOND RUDORFF, *THE VENICE PLOT*   *The*

overpriced, kitsch ridden tourist trap. It is also the most imaginative, awe inspiring place I have ever traveled to."

If Venice is to be saved from becoming little more than a tourist theme-park, a desperate need exists for strategic, integrated urban planning and draconian tourism management. Some would argue that it is the type of tourist that first needs to be changed if such planning and management are to be effective. Partying tourists to the annual Carnival and to big draw concerts in St Mark's Square should be curtailed in favour of cultural tourism. Sustainable tourism development needs to transform tourist activities and funnel the revenue into an economic base to finance the maintenance of Venice's infrastructure and its artistic and environmental capital. Reviewers of the city's strategic action plan for sustainable development, however, were none too optimistic in their conclusion that

"there is considerable concern among the assessment team that the economic promotion of…tourism is unsustainable and may result in progressive long-term damage to the Venetian environment." (PRESUD, 2003, 2004)

AND SO IT GOES ON…

IN September 2008, after years of delay and cost overruns of millions of euros, a new bridge opened linking the rail station with the car-park, which city officials had deemed "vital for improving the flow of tourists". For many of those already mistrustful of the claims made by the MOSE Consortium and sceptical about their technical capabilities, the bridge became the new hallmark example of Venetian engineers as Keystone Cops. The bridge had actually been erected and put in place for more than a year but was closed to pedestrians during this time as the engineers scrambled to strengthen the foundations and improve its functionality. Even so, the finished and finally opened ultramodern steel and glass structure controversially fails to accommodate the disabled, its railings burn the hands of pedestrians during hot summer days, and it is feared that its steps of variable height will trip people up and send them to hospital with sprained ankles. Many Venetians regard such a bridge that caters to tourists (saving them a ten-minute walk) as a completely

unneeded and grotesque waste of money that could have been better applied toward subsidised housing for locals. Tellingly, the grand opening celebration was ultimately cancelled in fear of demonstrating protesters whose opinions, as always, seem to have been ignored. MOSE, both redux and presaged, perhaps?

"If they can't build a simple pedestrian bridge without all these problems, how can we trust them to build something as complicated as the flood barriers?",

more than one critic has asked.

And finally, in December 2008, came the event whose arrival everyone knew was only be a question of when, not if. Once again the Adriatic poured into the city to a height of 156 cm above normal, the largest flood in two decades. St Mark's Square was under more than a metre of water and the city paralysed because the speed of inundation meant that the raised wooden walkways or *passerelle* could not be deployed. And of course, this being Venice, the entire event became a passionately debated issue. Supporters of MOSE stated the obvious – that if the protesters had only stopped complaining and let them build the mobile barriers years ago, then none of this would have happened. Opponents countered, stating that were it not for the channels dug at the lagoon entrances to accommodate the barriers, the flooding would have in fact been 20 cm lower. The mayor, not a supporter of MOSE, conceded that the barriers would have protected the city from floods such as this one which, it was worth noting he added pointedly, were only bi-decadal in occurrence. Nonsense, retorted the construction Consortium spokesperson, who pointed out that the barriers are designed to deploy for substantially lower *acqua alta* events as well, and when looking back over the records, would have protected the city from fifty-five of such lower magnitude floods in the last decade alone. Then, the opponents mentioned that the absence of any serious damage resulting from this latest quarter-century flood suggests that MOSE is, just like the new bridge, unnecessary as well as being of course both unwanted and uneconomical. But then, the barrier proponents came back and stated that a set of classical manuscripts of world renown located in the national library had been "saved by a whisker" this time, so that of course the barriers are needed. And then….so on and so on…

Germans, it was rumoured, saw the law as something to be obeyed, unlike the Italians, who saw it as something first to be fathomed and then evaded. – DONNA LEON, *ACQUA ALTA*

"SHOULD I TAKE MY TIME OR GO SLOWLY, SLOWLY?"

CONTENTIONS AND DELAYS: The woeful history of environmental management in modern Venice has resembled a contact sport characterised by prolonged time outs in which wounded parties retreat back into their respective corners while public officials seem incapable of speeding up processes of decision-making and implementation. Optimists contend that all problems are manageable and that the best way to insure that the city is really not sinking as some t-shirts promote (1) is to embrace the MOSE flood barriers. Pessimists rankle that the review process has been flawed beyond belief through being mired in corruption and in ignoring the voices of the many who remain opposed to MOSE and whose protestations on banners (2) or graffiti (3) can be seen about the city, including even out on the construction site itself (4).

Colour versions of all uncropped images can be accessed at **www.libripublishing.co.uk/veniceland**.

The state of Venice held its vicious sway, corrupting alike the ruler and the ruled, by its mockery of those sacred principles which are alone founded in truth and natural justice. — JAMES FENIMORE

EDUCATION OUTREACH: Pavement heightening and infrastructure renewal projects carried out around the city by Insula engineers have been worthy examples in how to capitalise on construction site fencing to spread the message about Venice's environmental problems and remedies (1, 2). One can learn, for example, about the Herculean extent of the reparations (3) and the actual processes involved (4, 5, 6). The most extensive set of information boards were those posted near St Mark's Square which identified problem areas (7), provided historical reference to flooding incidence (8), explained the mechanisms in which high tides affected the area (9, 10, 11) and the interventions underway to mitigate these (12, 13), as well as the overall project goals (14). Elsewhere, interpretive signs provide rudimentary information about historical and natural attributes of the lagoon (15).

COOPER, *THE BRAVO*   *Once this city's destiny was guided by the most virtuous minds, slaves to unbending laws, to standards of behavior, to a community that respected citizen and foreigner*

123

Pulizia dei rii interrati

L'intervento di pulizia libera il condotto di rio terà ai Saloni e rio terà dei Catecumeni, ormai completamente ostruiti, asportando l'accumulo di fango e detriti, che oggi rappresentano un oggettivo pericolo per l'igiene e la salute pubblica.

Da un'apertura praticata nel volto del condotto sotterraneo attraverso la pavimentazione soprastante, delle speciali macchine azionano una ventola che "risucchia", convogliandolo in un serbatoio stagno, il sedimento fangoso rimosso da un "mini-digger" idraulico; dal serbatoio il fango viene trasportato direttamente alla bettolina ormeggiata in canale e da qui a una discarica controllata.

Cleaning of covered canals

The cleaning operation clears the Rio Terà ai Saloni and Rio Terà dei Catecumeni sewers, which are now completely obstructed, removing the accumulation of sludge and debris that are now a real danger to public health and sanitation.

From an opening made through the pavement into the vault of the underground conduit, special machines operate a fan that sucks up the muddy sediment that has been dislodged by a hydraulic mini-digger and takes it to a water-tight tank; from the tank the sludge is conveyed direct to a lighter moored in the canal and thence to a controlled dump.

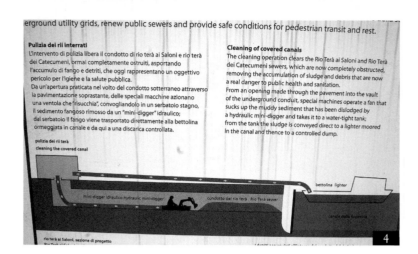

pulizia dei rii terà
cleaning the covered canal

mini-digger idraulico hydraulic minidigger
condotto del rio terà Rio Terà sewer
bettolina lighter
canale della Giudecca

rio terà ai Saloni, sezione di progetto
Rio Terà ... 4

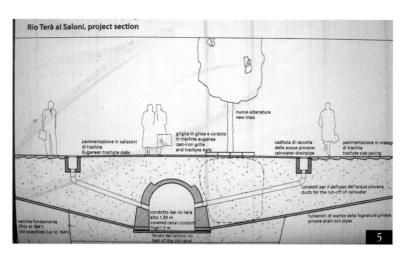

Rio Terà ai Saloni, project section

nuova alberatura
new tress

pavimentazione in salizzoni di trachite
Euganean trachyte slabs

griglia in ghisa e cordolo in trachite euganea
cast-iron grille and trachyte kerb

caditoia di raccolta delle acque piovane
rainwater drainpipe

pavimentazione in masegni di trachite
trachyte slab paving

vecchia fondamenta (fino al 1841)
old quaysides (up to 1841)

condotto del rio terà alto 1,50 m
covered canal conduit high 1.5 m

fondo dell'antico rio
bed of the old canal

condotti per il deflusso dell'acqua piovana
ducts for the run-off of rainwater

tubazioni di scarico delle fognature private
private drain soil pipes

5

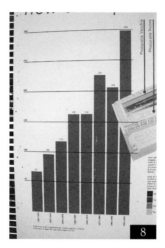

8

9

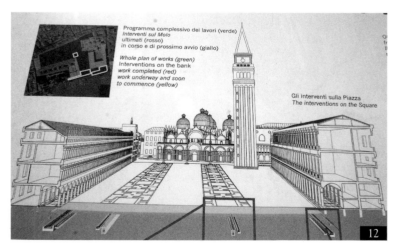

Programma complessivo dei lavori (verde)
Interventi sul Molo
ultimati (rosso)
in corso e di prossimo avvio (giallo)

Whole plan of works (green)
Interventions on the bank
work completed (red)
work underway and soon to commence (yellow)

Gli interventi sulla Piazza
The interventions on the Square

12

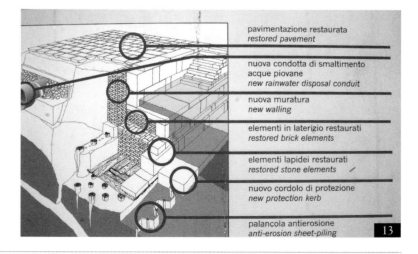

pavimentazione restaurata
restored pavement

nuova condotta di smaltimento acque piovane
new rainwater disposal conduit

nuova muratura
new walling

elementi in laterizio restaurati
restored brick elements

elementi lapidei restaurati
restored stone elements

nuovo cordolo di protezione
new protection kerb

palancola antierosione
anti-erosion sheet-piling

13

alike, that practiced tolerance. Now, like Italy, the city is corrupt to the core. – DAVID ADAMS CLEVELAND, WITH A GEM-LIKE FLAME   Corruption still crept in. Plots and pet projects flourished.

Defend the city from medium to high tides and wave motion

In order to allow pedestrian transit even in high water conditions, pavements will be raised (to the hight +120 cm) in the areas that are now lowest and thus subject to frequent flooding. The restoration of waterfronts, foundations and embankments will eliminate deterioration that has been aggravated by years without maintenance, and will preserve them in future from the degenerative action of wave motion.

Provide safe conditions for pedestrian transit and rest

The paving is now disconnected and uneven, with trachyte slabs laid and concrete and asphalt in the centre. The work will eliminate sagging and discontinuity in the pavement, thus optimising the possibility for rainwater and tide water to drain away. The design for the new pavement, which will be constructed making use of the old trachyte slabs, reproduces the shape of the old bridge situated at the confluence of the

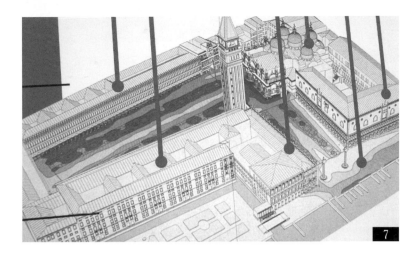

6

7

10

11

The objective of the interventions is to protect the area from the most frequent floods, to restore the pavement and to improve the condition of the area underground.

The project aims at protecting Saint Mark's Square from the most frequent high tides and will ensure the capacity to traverse the square at tides of up to +110 cm through approaches that respond to each of the three types of flooding: to combat water that surmounts the lagoon bank, the bank and the pavement behind it will be raised and restructured; to combat water that comes up through the drain pipes and from underground filtration, the interventions will restore and then close-off the ancient tunnels under the square and then will lay a new rainwater disposal network.

The collapse of these tunnels has, in turn, caused much of the deterioration of the pavement above. They will be restored without modifying the present level of the square. Furthermore, will be placed a layer of bentonite under the pavement of the square, in order to render it impermeable.

The interventions will include the re-systemisation of the underground infrastructure network, which will occur at the same time as the works for the defence of the square.

The local defence of the square is part of the system of measures to protect the city and lagoon from flooding and is integrated with the MOSE system under construction at the three lagoon inlets to defend the entire lagoon area from all high waters, including extreme events.

Gli in
La pr
(dal p

14

15

Webs of conspiracy were woven. – STEVE BERRY, *THE VENETIAN BETRAYAL*   *If the city survives at all, it will be as a theme park for rich tourists, as Veniceland, a wholly-owned Disney subsidiary*

125

FUTURE: Along the *Fondamente* Zatterie can be found one of the remnant suggestion boxes into which concerned citizens in days of the Republic would deposit their ideas about how to save the city from its, even then recognized, environmental problems (1). Once, on a whim, I placed a long list of prescriptions in this box and returned a year later to see it still lying there on the inside floor, ignored. The biggest question facing Venice concerns precisely who the city is to be saved for. Some international observers have suggested that the only hope is for the city to be completely surrendered to tourism and that some organisation such as Disney should run it as a theme park by charging admission. Disney is certainly interested as can be seen by their advertisement on the back cover of the prestigious *Harvard Magazine* (2), sent out to thousands of influential people around the world, and the fact that they have already established a beachhead in the city (3) near the Rialto Bridge. Meanwhile, megalomaniacal city officials and tourism planners contemplate modernising the city to make it more friendly for rapid transit (4), catering to what is expected to soon be more than twenty million tourists a year. Venice's future is therefore very much in doubt as a result of being loved to death, symbolised perfectly by a t-shirt logo depicting heart-shaped eye sockets on the skull atop the gondola *ferro*-shaped crossbones (5).

with actors dressed up as the Doge and the Council of Ten and catering by McDonalds. – MICHAEL DIBDEN, *DEAD LAGOON*   Everybody is talking about saving Venice…except most of we

Every great destination tells a story.
Write your family into the next chapter.

ADVENTURES
Disney

Venetians. — RAYMOND RUDORFF, *THE VENICE PLOT*   *One ought to be required to pass a test before being permitted to enter St Mark's.* — SALLY VICKERS, *MISS GARNET'S ANGEL*

127

"Venice is one of the few world heritage sites that perpetually runs the risk of vanishing completely."

Davis, R.C. and G.R. Marvin, *Venice, the Tourist Maze: A Cultural Critique of the World's Most Touristed City* (2004)

"Venice has adapted throughout its past. But if it is to survive through the 21st century, it will have to adapt like never before."

The History Channel. *Go Deep: Saving Venice* (2009)

Conclusion

Future of a Reinvented City

VENICE is closer to death than it ever has been. The A & E documentary concluded that Venice is like a sick patient, constantly revived but never cured. It has long been recognised, as summarised in *The Science of Saving Venice*, that the future health and survival of Venice and its lagoon depends on creating some form of balance among the needs of the environment, industry, agriculture, tourism, and Venetian citizens. The critical issue in all this is just exactly what *is* that balance point and exactly just for *whom* is the city being saved: the sixty thousand residents as a living and working urban landscape set in a sustainable environment, or the eighteen million tourists as a preserved fossil, a bauble set in amber?

The MOSE barriers, whose genesis is such an affront to the public participation process and possibly to financial responsibility as well, will almost certainly work to keep out the floodwaters and thus "save" Venice, at least for the foreseeable future. The city, however, is trapped in such a vortex of demographic decay that unless something is done to correct this, there will simply be nothing living left *to* save. So again, for whom is Venice being saved? Like Romulus and Remus, shameless city officials continue to suckle at the economic teats of the tourism wolf that is devouring their city through their plans to make Venice more "tourist friendly". So now we have ideas discussed for an underwater subway out to the Lido, an elevated monorail throughout the city, an underwater rail-line that would funnel tourists from the airport to Venice's Arsenale via Mestre and perhaps Murano, a network of moving pedestrian pathways snaking around the historic centre, etc. And always, hovering there in the background like an ominous spectre from a Bergman film, is the decades-old but as yet unacted upon urban renewal plan, sitting ready to be dusted and taken off the shelf from some bureaucrat's office, a plan that has a network of underwater roads throughout the lagoon and calls for the levelling of half the historic city in order to provide space to attract international resort developers who might want to build. It all simply and truly boggles the mind. And so again, is this *really* the sort of Venice that is worth saving?

If the existing Venice cannot be saved, can perhaps a new one be created to replace it? Julian Barnes, in his Booker Prize nominated novel *England, England*, satirises the creation of tourist "destinations", in his case a simulacrum on the Isle of Wight that contains all those best-loved elements of an imagined England: quaint pubs, bucolic landscapes, behaving royals etc. And in *The Second Venice (Venezia Due)*, self-published novelist Askin Ozcan posits a future where, due to restrictions put in place by the Italian Government to limit the number of tourists permitted to enter the real Venice in order to protect it from destruction, a new, better Venice is built right beside it on the mainland in order to cater to the not-to-be disappointed tourist hordes. This fake Venice, financed and managed by an American consortium led by a hamburger restaurant chain and a cola soft drink company, is a resort-city designed with artificial water features resembling canals in which gondolas attached to an underwater rail system move past chemically-aged buildings which are Hollywood-type facades. Other features include museums holding collections of forgeries, a new train station called Santa Cola rather than the Santa Lucia original, soft-drink dispensers and fast-food restaurants in every *campo*, and of course, an enormous underground parking garage to make visiting convenient.

A cute, blackly humorous and of course completely unbelievable little piece of satiric fluff you suppose. Truth, however, as we all know, can indeed be stranger than fiction. For in reality, it *is* to the mainland where many are looking to reinvent the Venice of the future….just as, it is alarming to remember, did their forefathers almost a century before when, infected by the same viral-mindset, they gave Venice, and therefore the world, the blight that is Porto Marghera.

The seminal *The Venice Report* concludes with the statement that as Venice no longer has funds to plan its own future, these decisions now lie in the hands of private investors. Given a city government and populace who seem impotent to decide their own fate, what big business wants, big business generally gets. In other words, according to the Report "it is private money that proposes and the Commune [city government] that bends to its will." And in a city that is not being run "with intelligence or dignity" such that "adequate urban planning is notable by its absence", a wild west mentality pervades in terms of development.

With the industrial area in rapid decline, a new "Venezia City" is planned to be built on the edge of the lagoon which will include a hundred-metre tower (that old phallic chestnut of

The Philistines are everywhere. A vulgarizing mob that has taken hold of this once great city – and they're not all foreigners either but Venetians themselves. – EDWARD SKLEPOWICH,

unimaginative architects), a business centre (business, in Venice?), parking lots (of course), and hotels (there can never be enough of these it seems). Then there is the planned "Tessara City" development to be built around the airport, which incidentally is run by a Berlusconi associate. Located here will be a huge hotel by "starchitect" Frank Geary (since Bilbao, every city believes that all it needs to reinvent and resurrect itself is another crumpled paper-type building), shopping centres (Benetton, after all, is a regional company), a new football stadium (despite having a team in the bottom league), and…wait for it…a Las Vegas style casino (one wonders, in a surreal nightmare of nesting Russian dolls, if this is to be modelled after the *Venetian Las Vegas* casino, which of course is itself modelled after the real Venice, which…)

Plans also exist for a new commercial port and transport hub to be constructed at Marghera as the petrochemical works are being downsized or closed and the government is looking for alternative employment. This new port would require deepening of the Malamocco inlet and navigation channels in the lagoon to enable entry of huge ships from the Adriatic. At one time it was forbidden to dig canals more than four metres deep. Now, "aquatic highways" of over twenty metres depth snake their way throughout the lagoon to accommodate boat traffic. Developers of the new port are not in the slightest bit worried by scientific evidence that suggests that such channel deepening would contribute to the further degradation of the lagoon by increasing tidal currents and wave energy. Why give credence to serious scientific investigations when the Port Authority states their confidence that the judicial use of MOSE, opening and closing the floodgates on demand, will rectify any hydraulic imbalances that might result from deepening the channels? Let's all put our faith in the floating basket of Moses, the techno-prophet who will lead us to the promised land of the new, improved "Venice on the mainland": the land of milk, honey, and other development-ensuing riches.

And so the Port Authority accuses the Venice in Peril Fund of "pseudo-scientific profiteering" and threatens to sue the UK-based organisation over their comments that warn about the inherent dangers of deep-dredging. At the same time they decline to participate in an open forum discussion about the scientific repercussions of the project. It should be mentioned that Venice in Peril, rather than being some Greenpeace-esque, tree-hugging organisation, actually came down in favor of the MOSE project after reviewing the objective scientific evidence. But again it appears that those Venetians with the authority to exact the greatest repercussions to the functioning of the lagoon, are not interested in countenancing any opinions that might force them into being more rigorous when it comes to adjudicating and defending their current pet project. It seems that the governing bodies have learnt absolutely nothing over the public relations fiasco of the last decade concerning the whole MOSE soap opera.

Noted author and Venetophile J. Morris once wrote that she thought that one solution to the problem of Venice was to let her sink, an opinion shared by Debray: "Let it slide into oblivion, all of it." Venice's future continues to be hotly debated, including an interesting exchange occurring in the international English language press in 2006. Berendt, for example, believed that the salvation of Venice, one of the world's most precious cities, was worth *any* price, even if that meant the managed theme-park option for the future. Another observer countered, stating that far too much money had already been wasted on restoring Venice. This was also the view held be another contributor to the debate who pessimistically concluded that Venice should be allowed to drift slowly to a stately death: "to sink back to the mirage upon which it has always floated" such that it can "live forever there in the fairytale land of the imagination" rather than suffer a fate worse than death through being turned into a modern Disney-fied fairyland exhibit. The supposition here of course is that Venice, if allowed to sink beneath the waves, would be unvisited by tourists. But there is no underestimating the magnetic strength of Venice's pull upon the tourist psyche. A far more likely scenario would be that pictured on apocalyptic posters available for purchase in kiosks around the city. Here, the Venice of the future is depicted as an underwater graveyard still visited by hordes of besotted tourists staring out from the porthole windows of their cruise-submarines at the submerged ruins…as well as, of course, at each other.

Now may be the last chance to save the real Venice, while there is still some reality left to be saved. It is as nowhere else on the planet, a place that has been sacrificed through being

DEATH IN A SERENE CITY   *It is Italy, our own Italy, that is responsible for the slow destruction of Venice, and that includes us Venetians…Venice has been declared*

ruthlessly exploited for tourism profit. And who is to blame for all this? Davis and Marvin pull no punches in unsympathetically stating that it is the self-destructive Venetians themselves who have sold out and continue to bring about the abject ruin of their once-marvellous city; the "Death of Venice Through Tourism" as they state it.

Unless some timely and drastically corrective action is taken, the future of this most remarkable and treasured of the world's cities, if we are to believe M. Plant's dire warning, looks bleak indeed. Paraphrasing her:

> "Meanwhile the seas [and tourism floods] are rising. In the city of the apocalypse, the four golden horses are at the ready, pawing at the porch of San Marco, waiting to haul the city out of the waters [and away from the tourists] and into the sky..." (Plant, 2002) [my additions]

And so it may be that Venice, the "miracle of the marshes", that one-time humble refuge, then emerging city-state, then prideful world empire, then decadent fleshpot, will live out her final days as the world's playground: "Veniceland", the Queen, no longer of the Adriatic, but of the Waterpark.

A second true story of remarkable prescience...

> Imagine a port city-state on the eastern side of the Italian peninsula that was the mercantile centre of its day through its dominance of trading along the length of the Mediterranean. A city that freely admitted settlers from other nations as long as they were interested in commerce. A city that minted its own currency and was ruled by a complicated oligarchy. A city whose citizens had a reputation for refined living that was so extravagant and so decadent that it became synonymous with pleasure and luxury and raised the jealous ire of its rivals. A city that contained the most magnificent dwellings, religious temples, administrative buildings, and works of art of its day. A city that became *the* essential place to visit. And finally imagine that this city's population had at one time been driven away due to malarial mosquitoes from the surrounding marshes and whose ultimate fate much later would be to be destroyed through being completely inundated by water. Imagine that.

Venice? you suppose. No. It was Sybaris, lost to the waters almost a millennium before the first refugees had ever entered the Venice lagoon. Look it up. "Almost everything you can think of has been done before," wrote de Combray about what he referred to as "the pending tragedy of Venice" resulting from "the recurring sensation the city has already sunk spiritually". Ouroboros. Meanwhile, Amphitrite awaits patiently her prize...

a playground. Anything serious must be taken away to clear space for tourist shops, restaurants, hotels, that sort of thing. – JANE TURNER RYLANDS, *VENETIAN STORIES*

133

A candid conversation
with the author

How do you think this book will be received?

One of the ironies of Venice is that, whereas eighteen million tourists stream through each year, when it comes to scholarship there is really a small-town protectiveness and defensiveness that characterises opinions. I have experienced such attitudes before, as for example when I organised a big international conference at Harvard University concerned with restoring the Mesopotamian marshes of southern Iraq that had been destroyed by Saddam Hussein. There, amongst the donor countries involved in the work, the British were critical of the Americans, who in turn were critical of the Japanese, who were critical of the Canadians, etc., with the Iraqis having little time for any such petulant infighting. Because shades of this same sort of territoriality pervade all restoration work in Venice undertaken by foreigners, and because even amongst this group I am an outside observer (notwithstanding having lived and taught there for three summers), I expect some may be critical of my efforts. However, I believe that the strength of the more objective, outsider viewpoint will offset any nuances that I may have missed, simplified or perhaps reported incorrectly in my overview in these pages. Beneficial views are often best obtained from a distant perspective. In contrast, frequently I have sat in meetings in Venice wherein one group of closely involved and otherwise well-meaning individuals has seemed to be blindly and unreasonably critical of those from another such group despite all working toward what, to outside eyes, appears to be a common and compatible goal.

Then there is the issue about which I am very conscious: namely that despite being a one-time, working resident of Venice, I am a British and Canadian citizen. Venetians are always angry when other Italians are critical of their city, just as Italians are of northern (Anglo-Saxon) European or other, in their minds, meddling, internationals who are critical of what they, the Italians, consider to be an issue of domestic sovereignty. Interestingly, often the harshest criticism about any book on Venice written by an outsider (no matter how perceptive or well written these may be, as for example, works by J. Berendt or J. Martin), come from those onetime foreigners who now live in Venice and who can cavalierly dismiss such a book simply on the basis that no matter how good the research might be, because the author either does not live there fulltime, nor speaks fluent Italian, much less *Veneziano*, then the book must have little merit. (Venice is really a quite a funny place where affectations still pervade society to a degree matched by few other locales. It is, for example, the only place I have been where it is possible to be at a social gathering and see expat Brits decked out in cravats and white shoes or elbow gloves and cigarette holders, seemingly as if they had just walked off the set from the latest Hercule Poirot film).

Then there are the ardent Venetophiles, those whose near obsession with all things Venice enables them to dissect every movie or novel about Venice, frame/page by frame/page, to see if it were possible, for example, for the protagonist to be able to get from one location to the next in the amount of time implied by the film/book's narrative (recently, for example, questions have been raised about how it is possible for Angelina Jolie and Johnny Depp in *The Tourist* to walk out onto the balcony from their room in the Hotel Danieli on the Riva degli Schiavoni and be able to gaze out at the Ponte Rialto, much less to have one disembark by boat at the Venice airport while the other drives off by water-taxi immediately into the Guideca Canal). My defence is that the present book is very much a *primer*, condensing and reinterpreting the copious work from a legion of dedicated scholars who conducted the primary research. And as with any such synthesising compilation, it is certain that mistakes and omissions do exist. And given the predilection of Venetophiles to point out such mistakes, I expect to receive more than one letter of correction in this regard.

So in the end, I think that if any critic is looking for errors in a few of the details, s/he may very well find them. I only hope that I have done justice to the complexity of the entirety, this being the first holistic account of Venice's myriad environmental *and* social problems.

You paint a dire picture for the environmental future of Venice and its lagoon. How do these problems compare with those experienced by other areas?

Although certainly subjected to being severely poked and prodded, the good news is that the Venice Lagoon is not in danger of being destroyed any time soon. And of course this is something that cannot be said of several other equally prominent and unique water bodies around the world such as Lake Chad, the Aral Sea, the Mesopotamian marshes, the Dead Sea, the Azraq Oasis, the Tonle Sap Great Lake, etc., the very existence of some of which into the next century seems doubtful. So in this regard, the future of the Venice Lagoon looks comparatively favourable. However, when one examines other important wetland systems within the developed world, such a conclusion may be overly optimistic. Within Europe, for example, the Norfolk Broads and the Carmargue are both in much better shape than the Venice Lagoon in terms of their present ecosystem integrity as well as having in place management systems to preserve and, if necessary, restore that integrity should the need arise. Even within Italy there are worlds of difference in the effective way in which everything from environmental protection to ecotourism operates in the nearby Po Delta lagoon and marshes compared with the Venice Lagoon. And in North America, the management of the ecocultural wetlandscapes of the Everglades and the Bay of Fundy marshlands, for example (both nominated as new natural wonders of the world), eclipse anything currently in place in the Venice Lagoon. And speaking of Venices, whereas the world's second-most tourist city, Las Vegas, shamelessly offers up a simulacrum of the real Venice in the form of one of its casinos, the remarkable way in which that American city comprehensively planned and now manages its own nearby wetlands (as I have written about in *Landscape Restoration Design for Recreation and Ecotourism*) is a model that the real Venice would do well reciprocally to copy. Then there is the wonderful public education program about the Ballona Wetlands in Los Angeles, located near Venice Beach (see my edited volume *Facilitating Watershed Management: Fostering Awareness and Stewardship*) which again could serve as a model for how the original Venice could and *should* approach its own environmental education. And finally there is the meritorious example of Boston, (to whose nineteenth-century Venetophiles such as John Singer Sargent and Henry James we owe much for the way we appreciate Venice) where the legacy of that sister-city infatuation is still present today in the form of Isabella Stewart Gardner's Venetian-styled palazzo-home-now-museum – with its Titian and Singer Sargent paintings, etc. – and real Venetian gondolas that can be hired on the Charles River (for a fraction of the cost of those back in Venice). Indeed, Boston's Charles River Watershed Association is truly one of the world's most successful such entities (as described in my *Introduction to Watershed Development: Understanding and Managing the Impacts of Sprawl*).

In short, there are lots of Venetian-affiliated locations around the world that do a much better job of preserving and restoring ecosystem integrity that the original Venice. Here are two further examples. Personally I have worked on the sustainability and land-use plans for two world-class cities that interestingly are physically or conceptually linked to Venice while at the same time offering many valuable lessons for how the one-time Queen of the Adriatic might polish up and regain her now all-too-tattered throne. Firstly, Hangzhou is the city where Venetian Marco Polo ended up and today it and its famous aquatic jewel, the West Lake, is one of the most historic and tourist sites in all of China. As part of helping the City plan future development in order to accommodate the projected influx of millions of largely domestic tourists, we – Harvard University students and faculty – presented city administrators with a series of alternative scenarios under the banner of "nature and humanity in harmony", a title that would certainly be incongruous if not outright oxymoronic for Venice. Secondly, in today's world, the city-state that most closely matches the wealth, mercantile influence, and boundless riches of historic Venice and its Veneto region is Abu Dhabi and the United Arab Emirates. As part of an international, award-winning team, our group provided the Capital City with an urban framework vision in which to accommodate the enormous increase in population that is expected over the next several decades. And completely unlike the situation in Venice's overall city planning today, the vision for Abu Dhabi, as explained near the beginning of the final report, explicitly states "The environmental framework deserves special mention as it helped shape the Plan at all stages and was particularly influential throughout the integrative work. Urban growth, ecological stability and the potential for regeneration have been reconciled in this sensitive ecological context." Further, preliminary planning documents actually go out of their way specifically to mention Venice and its ambivalent (at best) relationship with its own surrounding islands as an example of exactly just what Abu Dhabi should avoid with respect to its own archipelago (contrasted with other more positive examples such as Boston or Auckland harbours which should be emulated).

It is worth noting that both these examples of pragmatic urban design and planning for Hangzhou and Abu Dhabi (how successful they will be of course will depend on how and to what degree our recommendations are actually implemented) are remarkably different from the more fanciful rococo-renderings offered in a recent international design competition for Venice and its lagoon as mentioned and referenced in the text of the present book. Sadly, it appears that for much of the international design world, Venice still suffers from Robert Thayer's critique of the profession in his seminal work, *Gray World, Green Heart: Technology, Nature, and the Sustainable Landscape*, wherein much of design remains "dominated by the creation of pleasant, illusory places which either give token service to environmental stewardship values, or ignore them altogether."

So the bottom line is that whereas architects will continue to salivate over the rich heritage of Venice's built environment, sustainability planners and urban ecologists will continue to look upon the recent history of planning for the city and its lagoon with derision. The encouraging news is that sustainability is *finally* being recognised by a dedicated cadre of independent-minded city planners in Venice who are now actively challenging the hitherto accepted mindset. However, their voices and influence still remain underappreciated and so far little acted upon.

You provide an overview of the strong opinions held by those either supportive or opposed to the MOSE project and use the expression "gate sitters" in reference to individuals whom are undecided. Which side of the floodgates do you fall on?

MOSE has been the most contentious project undertaken in Venice in several centuries. For the overview presented in this book, I did attempt to maintain a semblance of neutrality in my reportage of the various points of views of players on both sides of the debate, but as with most things MOSE, objectivity is easier invoked than realised. By examining MOSE through a lens of my own experience participating in watershed management, environmental restoration, and urban planning projects around the world, I am forced to adjudicate the flood barriers from three viewpoints: justification, process, and product, and how they respond to the triple foundations of sustainability in terms of fostering ecology, economy, and social capital.

I agree with almost all outside observers that some form of physical separation of the City of Venice from the Adriatic is ultimately the only way to ensure that Venice will be protected from the discomforts and ravages of *acque alte*, and that at least in the short-term (until rising sea levels catch up) MOSE will adequately accomplish this. As for the big rallying cry from the green community concerning the perceived threats to the lagoon, I remain unconvinced that the flood barriers will exact an irreparable toll relative to all the other impacts on the wetland environment. We have to realise that the lagoon is a completely artificial environment that, had it not been for human tinkering, would long since have disappeared through completely natural processes. Does this make the lagoon any less valuable? Of course not. But it does, or it should, provide a perspective that is more accommodating to human management. The dichotomisation with which we portray our world – nature there, culture here – is incorrect and extremely damaging. More than anything, environmental restoration is an undertaking that demonstrates the seamless blending of the two; i.e. human hands rebuilding and protecting what human minds have reimagined and decided is worth protecting. And MOSE will do this.

Should MOSE do this? That is quite another question altogether. Here I am very sympathetic to those many critics who believe the mobile gates to be an unnecessary waste of money. MOSE, at the present time, really does make little or no economic sense relative to the infrequent and minor floods that paralyse the city. It is easy to champion any of a number of more meritorious targets, the funding of which would have much better served the needs of the populace, than an investment in the mobile flood barriers. As mentioned above, eventually some form of separation between lagoon and sea will have to be put in place. In my mind, other, cheaper options that have been posited would have been just as effective at this. Of course, the ecology of the lagoon might or would probably change. However, I remain unconvinced that the new ecology that would become established would be in any way a substandard ecology or one that would be markedly different from that which might have developed naturally. It must be remembered that MOSE is the compromise solution to the intractable problem of needing to both protect the historic city and lagoon at the same time as not hindering the revenue-making activities of the port of Marghera. Disregard and do away with the latter, and the solutions to the former become many, and the need for the super-expensive MOSE vanishes.

Finally, regardless of the efficacy of the final product or its perceived need, it is the process through which the decisions were made that is ultimately that which is so damming about the entire flood barriers story. MOSE will long be studied as perhaps *the* hallmark example of how *not* to undertake complex environmental planning. And in this regard, as a scholar of and participant in many such projects around the world that specifically rely upon civic engagement for their successful implementation, I am in complete agreement with every harsh criticism that has been heaped on MOSE by the marginalised, disenfranchised citizens of Venice.

So in summary, I question the particular need for MOSE, condemn the manner in which it was brought to fruition, but now that it is a *fait accompli*, am not unduly worried about its operation.

The case is made that Venice is more in danger of sinking beneath a flood of tourists than a flood of waters. If MOSE can control the latter, what in your opinion can be done to effectively manage the former?

This is the issue that lies at the crux of saving Venice. Finally, the vestigial rumblings made over the years by academics and planners about the city's inexorable descent into some sort of tourist-hell has captured the public attention through generating a clarion call to do *something* concrete about the issue. In November 2010, as this book was being prepared for publication with the title already chosen, a group of local protestors gathered in Venice to hand out stylised Disneyesque tourist maps and free "entrance tickets" as they cut a fake ribbon stretched across the Piazzale Roma in a mock inaugural opening of the new theme park called, of all things, "Veniceland". All staged as a citizen provocation against city officials catering solely to tourists.

For years observers have suggested that Venice put in place some form of entrance tariff that would support the sustainability measures needed to preserve a vibrant, living city. The difficulty in applying such a measure has been two-fold. Firstly, there is the pragmatic problem. Just to whom exactly should such an entrance fee be applied? Certainly those few remaining residents would be exempt. But how about those from neighbouring Mestre who work in Venice? Or those from the surrounding Veneto region? How about all Italians? And if them, why not all EU members? (i.e. make those rich Americans and Japanese pay, not us Europeans), etc. Many are the conversations with tourism and planning officials through which I have sat, arguing around and around about who should have to pay for what. Is it fair to charge the hordes disembarking in their thousands from the cruise ships yet not charge the millions who arrive by train or plane? Is it reasonable to levy a bed tax on hotels in Venice when so many sleep in cruise ship berths or are day-trippers from the mainland where hotels are cheaper? etc.

And then there is the second, philosophical problem. Some critics have argued against imposition of such an entrance fee as it would do a disservice through reinforcing the idea that Venice is nothing more than a theme park. One doesn't pay a fee to enter into the Lake District National Park, for example, because it is a real place where real people live. Why should Venice be any different?

Well the difference is that the Venice of today really has more in common with other rarified shrines of rampant consumerism such as Dubai, Singapore, the US Virgin Islands, etc. than it does with the Lake District. And if a shopper expects to encounter entertainment while indulging in his/her hobby/addiction, why not charge admission some argue. We pay this willingly when we visit Disneyland, so why not Veniceland?

And so, getting back to the original question: What do I think of all this? Here are some answers in the form of further questions. First, what would happen if Venice and its lagoon were designated as a national park to enter which non-residents are charged admission and in which the locals need not fear that they would be treated as circus sideshow freaks? When one visits Banff National Park, for instance, one pays a fee. Yet thousands live there in the City of Banff surrounded by the wonderful scenery of the Canadian Rockies. Many are the environmental and building restrictions that come into play when an area obtains such protected status, as anyone knows who lives in the Lake District. Some have proposed that Venice would benefit greatly from such draconian restrictions. Certainly such spectacles as the large rock concerts in St. Mark's Square would disappear as would the egregious advertising now plaguing the city.

Second, what would happen if building ownership were tied to residency? In other words, one as a non-local is not permitted to buy property in the city unless one spends a certain amount of time there, paying for services that one needs in order to live? And if you have consequent difficulty in finding vegetables or a hardware store or a pharmacy, as you certainly will, then maybe you will support such with your tax euros. Other, rarefied and threatened places in the world have in place all kinds of restrictions hampering foreign or non-resident ownership.

But back to reality; where do things stand today? Finally, in October 2010, the Italian government drew up a draft plan that would allow

Venetian authorities to charge an entrance tax for all visitors arriving via air, train and cruise ship. At even only a meagre single euro per tourist such a fee would generate enormous revenue to allow the city to maintain its infrastructure, build subsidised housing for locals, keep schools open, etc. Good on paper; we will wait to see whether it will ever be implemented.

You certainly don't seem to have too high an opinion of the type of tourist visiting Venice. Is this entirely fair?

I'm afraid that it's more than fair. Einstein's blackly humorous observation that there are two things that are infinite: the universe and human stupidity, and that he wasn't sure about the former, is certainly apt when describing the bulk of tourists who visit Venice, as I think the anecdotes I present in this book clearly illustrate. Really, imagine disembarking from a ship and not even knowing if Venice is in Italy for goodness sakes! Where I *am* unfair is the implication that it is merely those tourists who visit Venice who are shallow. The reality is that mass tourism, regardless of where it is – and there is no polite way to say this – caters to the intellectually lowest common denominator. Ample evidence of this can be found on many internet sites where incredulous tourist guides take great joy in sharing stories about the egregious ignorance of their clientele. One of my own particular favorites concerns a tourist who was part of a large guided group walking the so-called "Freedom Trail" in downtown Boston, Massachusetts – a red line, painted on the sidewalks that conveniently links up half a dozen historic sites important for the American Revolution – and who was overheard to say: "No wonder we rebelled against the English, if they forced us walk along this stupid red line."

What *is* abjectly unfair is that far too many people have far more money than they have brain cells. For every hundred thousand mindless, middle-aged cruise ship passengers or vacuous, partying students who can afford to visit Venice, and where the charms and history of this most remarkable of cities is completely wasted upon them, there is that sensitive young dreamer from the Yorkshire moors or the Canadian prairies who may never have the financial wherewithal to be able to experience the city outside of her or his imagination or readings.

You present ample evidence from those who are highly critical of the capability of Italians in general and Venetians in particular for solving Venice's problems. In your opinion, is this harsh criticism justified?

The answer here is both a yes and a no: yes, if one examines in isolation their oft-times mismanaged and sometimes corrupt responses to solving Venice's problems (criticisms it must be stated that arise most forcefully from those most in the know, namely the Italians and Venetians themselves, and thus cannot be simply dismissed as being due to prejudicial northern Anglo-Saxon meddlers); but then no, if one assumes from this that such limitations are somehow the sole purview of Venetians and Italians. In short, when one reviews the evidence it is impossible *not* to be critical of how the locals have dealt and are continuing to deal with the socio-environmental landscape of Venice. On the other hand, it is wise to keep in mind the cautions about those living in glass houses.

It is easy to find examples of mismanagement and dishonesty elsewhere in the developed world, as for example, recently the inflated expenses of UK politicians and the economic problems of Greece and the other "PIIGS" EU nations; and who can forget the charade of the 2000 US election when ex-president Jimmy Carter volunteered his African democracy observers to help sort the mess out, etc. In fact, had this instead been a jeremiad about the US Gulf Coast, for example, almost all the criticisms of the Italian mismanagement of Venice cited here could quite easily be applied there as well: the fiasco in dealing with the aftermath of Hurricane Katrina, the rampant destruction of the protective coastal wetlands that exacerbated the disaster for the benefit of, among others, the oil industry, and more recently just how that oil industry in the form of BP and its partners paid the region back with an enormous spill that severely damaged locations such as…wait for it…Venice (in this case, that one located in Louisiana). Further, in the Preface I offer examples of environmental mismanagement by other nationalities that rival or exceed that of Venetians and Italians.

However, it must be admitted that there is one significant difference between the situation in Italy and that in many of these other examples, namely the inescapably greater role played by organised crime in the former. This is widely believed to be a defining and debilitating fact of life in modern Italy (Donna Leon's hugely popular novels, for example, must be unique in the genre of crime fiction in that although, as is customary protocol, the culprits are always identified by the end of the book, as often as not they escape justice). This has led some to suggest that all decision-making in Venice has been severely compromised as a result of these pervasive influences. But again even here it must be acknowledged that, in terms of a conflated system of governance and commerce built on rampant criminality and impunity, there are conditions today in other countries of the world that probably supersede anything untoward that goes on in either Venice or Rome.

Is public participation really the key to solving Venice's problems? And how important are locals to Venice after all?

It is certainly *a* key. This is a subject on which I elaborate at great length in my *Handbook of Regenerative Landscape Design*, particularly in relation to discussing such approaches to civic engagement as the Local Agenda 21 Protocol which originated at the 1992 Rio sustainability meeting, the aforementioned alternative futures scenario studies for such places as Hangzhou but also nearby Padua among other locales, the new emerging socio-scientific discipline of restoration design that balances the aspirations of people and nature, and the promising Vision for Venice framework from the International Institute for the Urban Environment in the Netherlands. In addition, in the foreword to that book, sustainability strategist Robert Abbott posits the *Imagine Calgary* case study as a useful model for Venice to adopt in terms of making its citizens visible.

All these examples are based on the supposition that there *are* people who live in Venice, and that it is important to hear, acknowledge, and then act upon their wishes for their city. But what happens if the point is reached when there are, or almost are, no longer any Venetian residents? At what point do the wishes of 18 million (and growing) visitors trump those of the 40,000 (and decreasing) locals? Some years ago the World Wildlife Fund made the difficult decision that, because of such low numbers, it was not in the best interests of the organisation in terms of their limited manpower and financial resources to spend any more time and effort trying to save a particular species of rhinoceros. At what point then, unless things turn around and trends are stopped or reversed, does someone make the same horrible decision here and say that there are, or effectively are, no more native Venetians who can be saved?

Before this particular endangered species of *Venezia habitus* is formally written off, however, it is worth noting that what makes certain landscapes truly memorable is that they genuinely are "ecocultural" constructs and amalgams. One's fond memories of the English Lake District, for example, owe just as much to the presence of Ambleside and Keswick as they do to Derwentwater and Windermere. And similar can be said for other areas where humans live in "natural" parks such as the Adirondacks in New York State or Banff in the Canadian Rockies. And so, once again I answer a question with another question: would Venice still be Venice if there were no local Venetians living there?

The word "future" is used in the title for this book. Can you describe what the Venice of the future will look like, say one hundred years from now?

As my Harvard colleague and renowned expert on modeling alternative futures of rapidly changing landscapes and cities, Carl Steinitz, likes to quip, quoting Nils Bohr: "Prediction is difficult. Especially when it's about the future." Regardless of whether one is a pessimist or an optimist, the Venice of the future will look remarkably different from what it was like, say twenty years ago, when some readers of this book may have had the fortune to be able to first visit the city at a time when it was still a place of magic and mystery.

In one extreme scenario, as the pointedly alarmist subtitle to this book has it, the future of Venice will be "bleak" indeed. Italy at that distant time will be bankrupt, due to even more widespread corruption and financial mismanagement than today, and Venice (as well as Rome) will need to be sacrificed to tourism to provide revenue for the entire state (whose new capital will be Milan). In consequence, Venice at the start of the next century will be operated as the world's most popular theme-park. Tourists, capped at about 50 million per year in order "to preserve the intimate experience" as the advertising will state it, will need to pre-book months or even years in advance and will be required to pay an exorbitant admission fee. However, it will always be possible to jump the queue with the appropriate amount of "baksheesh". The only residents of the city will be interpretive, costumed actors and hotel staff. Outside, there will be daily spectacles and evening shows. Inside, there will be more shows and 24-hour shopping. Venice of the future will resemble Las Vegas of today. The lagoon will be cleaner of pollution than it is today but it will be freshwater, the MOSE gates long since having outlived their usefulness and been replaced with the much cheaper option of permanent earthen-dam closure to the progressively rising Adriatic. The latter will be due to the perpetual failure of the United States and China and the reneging of India to sign international protocols to reduce greenhouse gas emissions. At certain times of the year, Venice will be empty of tourists and be used as a stage set for filming movies. Venetophiles, either those with sufficient funds to be able to gain admittance to the city or those many without such wherewithal, will still be able to savour their dreams of that bygone Venice of yore – the mysterious canals and crumbling palaces – through the ever popular industry of novels about the city. These will of course be perused on electronic readers, real books like this one having disappeared by the middle of the century. Equally popular, especially for the masses, will be high-tech video games in which one can vicariously experience the romanticism and allure of Venice as a historically layered palimpsest, lingering on whatever particular time period most attracts one's fancy. Venice in the future will become the Antarctica of today: a distant place of imagination that most can never afford to visit but whom are nevertheless happy has been preserved. In the future, Venice will no longer be regarded as the city-cum-museum that some now envision the place, but rather as a sort of city-cum-relic. And like all such relics, the withered yet miraculously preserved remains of the city will by then have assumed a quasi-religious aura far exceeding their true worth.

That is one vision of the future. Alternative, not so bleak, futures are many. In all cases these will necessitate the adoption of severe, almost draconian, measures to stem the bipartite flows of people: those of the millions arriving annually, but even more importantly, those of the hundreds emigrating every year. This latter, more than anything else, is the key problem facing the city today. For a Venice without residents *in situ* is, in the end, no longer really Venice at all. Instead, it is masqued imposter, an impoverished simulacrum called Veniceland.

Back to the Venice of today. A couple of personal questions. In terms of tourism, what is the best thing you can recommend doing in Venice? And what is the most unusual thing you yourself have done?

This sounds harsh, but the best thing to do in Venice is to leave it, at least during those mid-day hours when it is swamped by the flood of tourists. There are many gems throughout the lagoon to which one can flee the crowds, though even here one needs to be increasingly more imaginative, As recently as only several years ago, a visit to Torcello was a peaceful reprieve. Here one could see the remarkable Romanesque church, have a high-quality lunch, and wander about this quiet island of ghosts and nostalgia. Nowadays, it seems, however, that even sitting in the shade there is ruined by the din of boom-box toting tourists. So one must go elsewhere. The best place is the island of San Erasmo. Here one can spend a wonderful day enjoying a cacophony of birds while walking along empty coastlines and agricultural fields, canals, and wetlands. Gelato and drinks cost a fraction of what they do back in historic Venice. Shopkeepers are friendly and particularly interested in any tourist who would go to the trouble to visit their island. Dogs here are real dogs: the big, floppy-eared, happy rural kind, not the tiny manicured, tucked-under-the-arm while shopping, rat-dogs that have replaced cats in recent years in Venice. If one's ear is discriminating enough, it is even possible to discern the Venetian dialect among locals. And yes, these are *real* locals who actually live and farm here. Finally, it is even possible to find a spot secluded enough to throw off one's walking apparel and go for a skinny dip in the relatively clean waters in this part of the lagoon.

After time spent in the centre of Venice where water is the dominating presence, it becomes very hard, even amongst those who know better, to resist the temptation to take a Byronesque plunge into some canal to cool down from the often oppressive heat and humidity. And for that reason, most people, locals and tourists alike, head over to the Lido for a swim. This is the worst thing one can do on the Lido. The doggy public beach is filled with unsightly cigarette butts and middle-aged butts squeezed into Speedo swimsuits. Far better is to stroll about the forest in the incredibly atmospheric Jewish cemetery located on the island. One does not need to know anything about the history of Venice's ghetto, the world's first, to appreciate the beauty and serenity of the place.

With its omnipresent feelings of doom and death, Venice has always struck me as the oddest of places to go for a honeymoon or wedding reunion. Nevertheless, it is possible to have a romantic time in modern Venice, even in the summer. In the historic centre of Venice itself, the best thing to do is to break out of and escape from the so-called "Bermuda shorts triangle" area and simply go wandering about the still relatively quiet neighbourhoods that, against all odds, do still exist. Put away your maps and walk until you find a quiet canal and *ponte*. Sit down beside the hump-backed bridge and have a picnic while watching and becoming mesmerised by the reflections of light cast upward from the water onto the convenient screen made by the underside of the bridge. And if you've been fortunate to have found the right neighbourhood and are there at the right time of late afternoon or early evening, it is actually possible that you can enjoy yourselves in complete privacy for half an hour or so, the only sound being the lapping water at your feet and perhaps the faint music from a radio in someone's apartment above you, before the next local strolls past.

Having had the luxury of time as a result of living there, I've certainly done many unusual things in Venice, two perhaps more so in this regard than others. Once, in homage to the difficult passage made by more and more modern Venetians in terms of their leaving the historic city and moving to Mestre on the mainland, I walked there. The four-kilometre walk across the causeway was fine, as there is a sidewalk beside the rushing traffic, and with headphones on, I could almost forget about the trucks, cars, and tour buses, and enjoy the views of the lagoon. Once the mainland is reached, however, the verge disappears and one is forced to scramble through bush-filled ditches and squeeze underneath bridges within metres of the dangerous vehicular traffic. Having at that time recently survived a two-hundred-kilometre walking pilgrimage to Assisi, I was well familiar with the predatory zeal with which Italian drivers regard their pedestrian prey. So the walk from the lagoon edge to the Mestre train station was terrifying beyond belief.

On a much more positive note, the second most unusual thing I have done in Venice was also an homage. Friends and I collected some water from the canal outside my apartment and went to the cemetery island of San Michele. There we sought out and found the grave of Nobel Laureate Joseph Brodsky and in between readings of our favourite of his immortal words from his wonderful book about Venice, *Watermark*, we poured Venetian water over his gravesite, anointing as it were, his mortal remains.

With publication of this socio-environmental synthesis and summary, are you done with Venice?

For a scholar, Venice, even more so than Paris, to steal from Hemingway, *is* a movable feast. Once bitten into and savoured, it quickly and easily becomes incorporated into one's very body and spirit. And it seems that more than any other group, it is English speakers who are particularly vulnerable to succumbing to the lure and adding to the lore of Venice. More English-language books for instance have probably been written about Venice than any city excepting London and New York, which are the two most populous English-speaking cities. Think about that for a moment and one begins to realise just how odd it really is; i.e. that the third-most popular city in our mother prose is in a country where English is not the native language.

While living in Venice I restricted myself to reading only books about the city and was soon struck by just how good a job many novelists were doing at portraying both the manifest problems and the potential solutions. The results of some of that survey are included as the running foot-quotes in the present book. However, those quotations are only a small fraction of the many, some that I collected extending in length for several paragraphs. As such, at some time in the future I would like to return to those readings and more thoroughly investigate just how fiction has portrayed the socio-environmental milieu of the city. And in a related literary vein, I am also intrigued by the role that Venice has played in the modern crime novel and would like to explore that more as well. If one adds up the body count from more than a century of such crimson prose I'm sure that there is simply not enough room within the city's canals to stockpile all those unfortunate victims of this literary imagination.

References

Abbott, R.M. 2008. Visible cities: A meditation on civic engagement for urban sustainability and landscape regeneration. Foreword *in* France, R.L. (Ed.) *Handbook of landscape regenerative design*. CRC Press.

Ackroyd, P. 2009. Venice: *Pure city*. Chatto and Windus.

Albano, C., P. Frank and A. Giacomelli. 2007. Innovative design approach for the Fusina treatment wetland. Abstract, IWA *Internat. Conf. 2007*, Padova.

Alberni, A., A. Longo, S. Tonin, F. Trembetta and M. Turvani. 2006. Developer preferences for brownfield policies. *In* Alberni, A., P. Rosato and M. Turvani. *Valuing complex natural resource systems: The case of the lagoon of Venice*. Edward Elgar Publ.

Alberni, A., P. Rosato and M. Turvani. 2006. *Valuing complex natural resource systems: The case of the lagoon of Venice*. Edward Elgar Publ.

Ammerman, A.J. and C.E. McClennen. 2000. Urban ecology: Saving Venice. *Science* 289:1301–1302.

Anon. 2007. Controversy dogs Venice's first new bridge in 70 years. *The Independent*, Jan. 17.

Anon. 2007. In Venice, turmoil over a new bridge. *Internat. Herald Tribune*, Aug. 12.

Anon. 2008. *Venice review: Evidence*. Unpubl. document.

Anon. 2008. Venice opens a water bus for residents, minus tourists. *The New York Times*, Jan. 22

Anon. 2008. Protest over advertising in St Mark's Square, Venice., Feb. *The Art Newspaper*

Anon. 2008. Venice cancels opening ceremony for latest Santiago Calatrava bridge. *The Times*, Aug. 28.

Anon. 2008. Bridge trips tourists. *The Guardian*, Sept. 29.

Anon. 2008. Tourists warned to stay away. www.news.sky.com. Dec. 1.

Anon. 2008. Floodwater begins to recede in Venice. *The New York Times*, Dec. 2.

Anon. 2009. Venice versus the sea. *National Geographic Magazine*, Aug. 108–114.

Anon. 2009. Mock funeral for Venice dramatizes flight of residents from city's heart. *The New York Times*, Nov. 11.

Anon. 2009. Venice Port Authority defends dredging. *Maritime Journal,* Nov. 22.

Anon. 2009. Venice seaport to build power plant fuelled by algae. *Reuters News Service*.

Arts and Entertainment. 1996. *Miraculous canals of Venice*. DVD. A & E Ancient Mysteries.

Barbaro, P. 2001. *Venice revealed: An intimate portrait. Souvenir Press*.

Barnes, J. 1998. *England, England*, Vintage International.

Barr, D. 2006. Not fading, but drowning. *Country Life*, Nov.

BBC News. 2009. Is Venice dying? An interview with Francesco da Mosto and Donna Leon. *BBC* Nov. 14.

Berendt, J. 2005. *City of falling angels*. The Penguin Press.

Berendt, J. 2006. Venice; A city beyond price. *The Times*, June 10.

Bisa, S. 2008. Lagoon materials, or, a brief history of the lagoon of Venice as urban space. *In* Gili, M., M. Puente and A. Puyuelo. (Ed.s) 2008. Concurso 2G competition: Venice lagoon park. 2G Dossier.

Blewitt, J. (Ed.) 2007. *Community, empowerment and sustainable development.* Green Books.

Boele, N. et al. 2007. *Strategic objectives for Venice. Part A and B.* www.urban.ul.

Brodsky, J. 1992. *Watermark.* Farrar, Straus & Giroux.

Brown, P.F. 1996. *Venice and antiquity: The Venetian sense of the past.* Yale Univ. Press.

Buckley, J. 2004. *The Rough guide to Venice and the Veneto.* Rough Guides.

Campbell, M. (Dir.) 2006. *Casino royale.* Film.

Campbell-Johnston, R. 2006. If you love Venice, let her die. *The Times,* June 5

Campostrini, P. 2006. *Scientific research and safeguarding of Venice. Vol. V.* CORILA.

Campostrini, P. 2006. *Scientific research and safeguarding of Venice. Vol. VI.* CORILA.

Chang, C.Y. 2006. *A local emergence of cultural ecotourism: Westergasfabiek and Lazzaretto Nuovo.* Harvard Design School.

Ciriacone, S. 2006. *Building on water: Venice, Holland and the construction of the European landscape in the early modern times.* Berghahn Books.

CELI (Collegio di Esperti di Livello Internazionale). 1998. *Report on the mobile gates project for the tidal flow regulation at the Venice Lagoon inlets.* Regional Press Venice.

Clarke, H. 2006. Tide of opinion turning against Venice dam. *The Daily Telegraph,* January 29.

Consorzio Venezia Nuovo. 2007-10. www.salve.it

Consortium for Coordination of Research Activities Concerning the Venice Lagoon System. 2008-10. www.corila.it

Cotton, J. 2009. Fictional Venice. www.fictionalcities.co.uk

Crouzet-Pavan, E. 2005. *Venice triumphant: The horizons of a myth.* John Hopkins Univ. Press.

Da Mosto, F. 2004. *Francesco's Venice.* BBC films and books.

Da Mosto, J. et al. 2009. *The Venice Report: Demography, tourism, financing and change of use of buildings.* Venice in Peril.

Davis, R.C. and G.R. Marvin. 2004. *Venice, the tourist maze: A cultural critique of the world's most touristed city.* Univ. Calif. Press.

Debray, R. 1999. *Against Venice.* North Atlantic Books.

de Combray, R. 1975. *Venice, frail barrier: Portrait of a disappearing city.* Doubleday & Company, Inc.

Discovery Channel. 2008. *Venice, code red.* Megabuilders documentary, The Discovery Channel.

Fay, S. and P.K. Knightly. 1976. *The death of Venice.* Andre Deutsch Publ.

Fletcher, C. and J. Da Mosto. 2004. *The science of saving Venice.* Umberto Allemandi & C.

Fletcher, C.A. and T. Spencer. (Eds.) 2005. *Flooding and environmental challenges for Venice and its lagoon: State of knowledge.* Cambridge Univ. Press.

Foltain, J. 2009. Venice 'ruined by ads'. *The Times,* Aug. 2.

Forum per la Laguna. 2007-9. *Urban forum for sustainable development.* www.forumlagunavenezia.org.

France, R.L. 2002. *Handbook of water sensitive planning and design.* Lewis Publ.

France, R.L. 2003. *Wetland design: Principles and practices for landscape architects and land-use planners.* W.W. Norton.

France, R.L. (Ed.) 2005. *Facilitating watershed management: Fostering awareness and stewardship.* Rowman & Littlefield.

France, R.L. 2006. *Introduction to watershed development: Understanding and managing the impacts of sprawl.* Rowman & Littlefield.

France, R.L. (Ed.) 2007. *Wetlands of mass destruction: Ancient presage for contemporary ecocide in southern Iraq.* Green Frigate Books.

France, R.L. (Ed.) 2008. *Healing natures, repairing relationships: New perspectives on restoring ecological spaces and consciousness.* Green Frigate Books.

France, R.L. (Ed.) 2008. *Handbook of regenerative landscape design.* CRC Press.

France, R.L. 2008. Reparative paradigms: Sociological lessons for Venice from regenerative landscape design. *In* France, R.L. (Ed.) *Handbook of regenerative landscape design.* CRC Press.

Gili, M., M. Puente and A. Puyuelo. (Ed.s) 2008. *Concurso 2G competition: Venice Lagoon Park.* 2G Dossier.

Gordon, S. 2009. Venice holds mock funeral to mourn shrinking population due to rampant tourism. *The Daily Mail,* Nov. 16.

History Channel. 2009. *Go deep: Saving Venice.* Documentary The History Channel.

Hooper, J. 2006. Population decline set to turn Venice into Italy's Disneyland. *The Guardian,* Aug. 26.

Hooper, J. 2008. Venice flood fails to damp down fight over sea walls. *The Guardian,* Dec. 6.

Hubert, L. 1994. Western Europe and public corruption: Expert views on attention, extent and strategies. *Eur. J. Criminal Policy and Res.* 3:7–70.

IIUC. 2007-9. *Developing the economy from within.* www.urban.nl.P_Dew.

Insula. 2008-10. www.insula.it

Istituto Veneto di Scienze Lettere ed Arti. 2008-10. www.istitutoveneto.it

Jones, T. 2009. *The dark heart of Italy.* Faber and Faber.

Kay, J. 2006. Venice's real problem is organization and management. *The Art Newspaper,* July.

Kay, J. 2006. The magic Kingdom could save Venice. *The Financial Times,* June 13.

Kay, J. 2008. Welcome to Venice, the theme park. *The Times,* Jan. 3.

Kay, J. 2009. Venice is a management challenge. *Venice in Peril Newsletter.* Winter 08/09.

Keahey, J. 2002. *Venice against the sea: A city besieged.* St. Martin's Press.

Lasserre, P. and A. Marzollo. (Eds.) 2000. *The Venice lagoon ecosystem: Impacts and interactions between land and sea.* The Partheon Publ. Group.

Lauritzen, P. 1986. *Venice preserved.* Michael Joseph Ltd.

Lovric, M. (Ed.) 2003. Venice: *Tales of the City.* Abacus.

Magistrato alle Acque di Venezia. 2008-10. www.magisacque.it

Mambro, A. di (Ed.) 2006. *Il Parco di San Giuliano.* Communitas, Inc.

Martin, J. 2007. *No vulgar hotel: The desire and pursuit of Venice.* W.W. Norton.

McCarthy, M. 1963. *Venice observed.* Harcourt inc.

McCourry, C. (Director) 2005. *Venice: Tides of change.* DVD McCourry Films.

McGrouty, P. 2008. Venice to ban pigeon feeders from St. Mark's Square. *Spiegel On-line*, Dec., 2.

MIPIM. 2006. *Venice at the MIPIM 2006 project: Aims, local operators involved, and promoted projects.* www.forumlagunavenezia.org

MITVWACVN (Ministry of Transport, Venice Water Authority, Consorzio Venezia Nuova). 2006. *Measures for safeguarding of Venice and the lagoon.* Information poster.

MITVWACVN (Ministry of Transport, Venice Water Authority, Consorzio Venezia Nuova). 2006. *Venice. Mobile barriers at the inlets to regulate tides in the lagoon.* Information brochure.

Moltedo, G. (Ed.) 2007. *Welcome to Venice: Replicas, imitations and dreams of an Italian city.* Consorzio Venezia Nuova.

Morris, J. 1960. *Venice.* Faber.

Morris, R. 2009. Venice is experiencing a new economic crisis. *Spectator*, June 10.

Musu, I. (Ed.) 2001. *Sustainable Venice: Suggestions for the future.* Kluwer Academic Publ.

Newman, C. 2009. Vanishing Venice. *National Geographic Magazine*, Aug.: 88-107.

Nova. 2002. *Sinking city of Venice.* Nova Documentaries. PBS.

Owen, R. 2008. Venice cancels opening for hated Santiago Calatrava Bridge. *The Times*, Aug. 28.

Ozcan, O. 2005. *The Second Venice.* Outskirts Press.

Pemble, J. 1995. *Venice rediscovered.* Clarendon Press.

Pertot, G. 2004. *Venice: Extraordinary maintenance.* Paul Holberton Publ.

Piazzano, P. 2000. Venice: Duels over troubled waters. www.unesco.org/courier

Plant, M. 2002. *Venice: Fragile city 1797-1997.* Yale Univ. Press.

PRESUD. 2002. *Peer review for European sustainable urban development.* www.commune.venezia.it

PRESUD. 2003. *The city of Venice smart action plan. Draft.* www.commune.venezia.it

PRESUD. 2004. *Venice review. Evidence.* www.commune.venezia.it

PRESUD. 2004. *Peer review for European sustainable urban development. Performance assessment.* Draft. www.commune.venezia.it

Resini, D. 2006. *Venice: The Grand Canal.* Vianello Libri.

Rosato, P., C. Giupponi, M. Breil and A. Fassio. 2006. Evaluation of urban improvement on the islands of the Venice Lagoon: A spatially-distributed hedonic-hierarchical approach. *In* Alberni, A., P. Rosato and M. Turvani. *Valuing complex natural resource systems: The case of the lagoon of Venice.* Edward Elgar Publ.

Ruskin, J. 1960. *The stones of Venice.* Da Capo Press.

Ruskin, J. 2008. *The seven lamps of architecture.* Grove Press.

Rylands, J. 2005. *Across the bridge of sighs: More Venetian stories.* Pantheon.

Sabaeadv, J. 1987. Twenty years of restoration in Venice. *Arch. Di Venezia.*

Scearce, C. 2006. Venice and the environmental hazards of coastal cities. *Discovery Guides.* www.csa.com/discoveryguides/venice

Sciama, L. 2005. *A Venetian island: Environment, history and change in Burano.* Berghahn.

Spelman, E. V. 2003. *Repair: The impulse to restore in a fragile world.* Beacon Press.

Steinmann, M. 1971. A lost city discovered: Space-age instruments unlock the secret of Sybaris. In *Nature/Science Annual 1970.* TimeLife Books.

Tobias, A. 2005. *The dark heart of Italy.* North Point Press.

Tonin, S. 2006. What is the value of brownfields? A review of possible approaches, *In* Alberni, A., P. Rosato and M. Turvani. *Valuing complex natural resource systems: The case of the lagoon of Venice.* Edward Elgar Publ.

Transparency International (2010) www.transparency.org

Trembetta, F. and M. Turvani. 2006. Governing environmental restoration: institutions and industrial site clean-ups. *In* Alberni, A., P. Rosato and M. Turvani. *Valuing complex natural resource systems: The case of the lagoon of Venice.* Edward Elgar Publ.

UNESCO. 1979. *Venice restored.* UNESCO Publ.

Venice in Peril. 2007-10. *News articles.* www.veniceinperil.org/news_articles

Veniceword 2010. www.veniceword.com/mosesystem

Visconti, L. (Dir.) 1971. *Death in Venice.* Film.

www.wikipedia.org/wiki/corruption_perception_index

Willan, P. 2009. Anger at Coca-Cola deal to place vending machines in Venice. *The Telegraph*, Feb. 24.

Willey, D. 2009. Row over Venice Coke sponsorship. *BBC News*, Feb. 23.

Woodsworth, N. 2008. *The liquid continent – A Mediterranean Trilogy: Vol. II Venice.* Haus.

SOURCES FOR UNACCREDITED TEXT QUOTATIONS

Preface

"Venetophiles" (Martin 2007)

Chapter 1

"devastation" (Pertot 2004)

"missing bases" (Pemble 1995)

"pedestrianisation" (Pertot 2004)

"a bath of poison" (Plant 2002)

"What's the problem…" (Keahey 2002)

"Venetians have been…" (Fletcher and Da Mosto 2004)

"romance of decay" (Brown 1996)

Chapter 2

"save" (Pemble 2005)

"Venice is above all…" (Pertot 2004)

"diffuse city" (Fletcher and Spencer 2005)

"The Guardian of the Water…" (Fay and Knightly 1976)

"start at the source", "end-of-the-pipe", "dilution is the solution…" (France 2006)

Chapter 3

"Veniceland" (Dibden 1994)

"portrayed not as a present…" (Pertot 2004)

"bite and run" (Keahey 2002)

Chapter 4

"recreating instead of repairing", "rescued", "de-restoration" (Pertot 2004)

"tainted by…" (Musu 2001)

"a terrible waste…" (Lauritzen 1986)

"bureaucratic lethargy" (Keahey 2002)

"woeful Italian tradition" (Pertot 2004)

"mysterious", "secretive", "suspiciously complex", "defies forensic interpretation" (Jones 2009)

"Our overall conclusion…" (Fletcher and Spencer 2005)

"urban void" (Bisa 2008)

"to sink back…" (Campbell-Johnston 2006)

SOURCES FOR FOOT QUOTATIONS

Alison, Jane. *The Marriage of the Sea*. Farrar, Strauss Giroux (2003).

Begley, Louis. *Mistler's Exit*. Fawcett Books (1998).

Berry, Steve. *The Venetian Betrayal*. Ballantine Books (2007).

Brodkey, Harold. *Profane Friendship*. Mercury House (1994)

Charles, David. *Daughters of the Doge*. Pan Books (2007).

Cleveland, David Adams. *With a Gemlike Flame*. Carroll & Graf (2001).

Cooper, James Fenimore. *The Bravo*. Bibliobazaar (2006).

Coover, Robert. *Pinocchio in Venice*. Grove Press (1991).

Delalande, Arnaud. *The Dante Trap*. Phoenix (2006).

Dibden, Michael. *Dead Lagoon*. Vintage Crime (1994).

Dunant, Sarah. *In the Company of the Courtesan*. Virago Press (2006).

Elegant, Robert. *Bianca: A Novel of Venice*. St. Martin's Press (1993).

Giradi, Robert. *Vaporetto 13*. Delta (1997).

Goldman, William. *The Silent Gondoliers*. Del Rey (1983).

Hartley, L.P. *Simonetta Perkins*. Hesperus Classics (2004).

Healey, Ben, *Midnight Ferry to Venice*. Walker and Comp. (1981).

Hewson, David. *Lucifer's Shadow*. Delta (2004).

Hill, Reginald. *Another Death in Venice*. Signet (1976).

James, Henry. *The Aspern Papers*. Dover (2001).

Johnson, Velda. *Masquerade in Venice*. Dodd, Mead and Comp. (1973).

Jong, Erica. *Serenessima: A Novel of Venice*. Houghton Mifflin (1987).

Kanon, Joseph. *Alibi*. Time Warner Books (2005).

Laker, Rosalind. *The Venetian Mask*. Three Rivers Press (1992).

Langton, Jane. *The Thief of Venice*. Penguin (1999).

Leon, Donna. *A Sea of Troubles*. Penguin (2005).

Leon, Donna. *Doctored Evidence*. Penguin (2004).

Leon, Donna. *Blood From a Stone*. Penguin (2006).

Leon, Donna. *Acqua Alta*. Pan Books (1996).

Leon, Donna. *Friends in High Places*. Penguin (1996).

Leon, Donna. *Death in a Strange Country*. Penguin (1993).

Lovric, Michelle. *The Floating Book*. Regan Books (2003).

Lovric, M.R. *Carnevale*. Vintage (2001).

Martines, Lauro. *Loredana: A Venetian Tale*. Thomas Dunne Books (2004).

McEwan, Ian. *The Comfort of Strangers*. Picador (1981).

Meyer, Kai. *The Water Mirror*. Margaret K. McElderry Books (2005).

Myers, Beverle Graves. *Painted Veil*. Poisoned Pen Press (2005).

Ozcan, Oskin. *The Second Venice*. Outskirts Press (2005).

Pears, Iain. *The Titian Conspiracy*. Berkeley Prime Crime (2002).

Phillips, Caryl. *The Nature of Blood*. Vintage (1997).

Riviere, William. *By the Grand Canal*. Grove Press (2004).

Riviere, William. *A Venetian Theory of Heaven*. Hodder and Stoughton (1992).

Rudorff, Raymond. *The Venice Plot*. Berkley Publishing (1976).

Rylands, Jane Turner. *Venetian Stories*. Anchor Books (2003).

Schnitzler, Arthur. *Casanova's Return to Venice*. Pushkin Press (1998).

Sklepowich, Edward. *Deadly to the Sight*. Thomas Dunne Books (2002).

Sklepowich, Edward. *The Last Gondola*. Thomas Dunne Books (2003).

Sklepowich, Edward. *Death in a Serene City*. Avon Books (1990).

Sklepowich, Edward. *Liquid Desires*. Avon Books (1993).

Sklepowich, Edward. *The Veils of Venice*. Severn House (2009).

Spark, Muriel. *Territorial Rights*. Perigree Books (1984).

St. Aubin de Teran, Lisa. *The Palace*. HarperCollins (1997).

Sterling, Thomas. *Murder in Venice*. Dell Publishing (1955).

Thompson, David. *The Mirrormaker*. HarperCollins (2003).

Unsworth, Barry. *The Stone Virgin*. W.W. Norton (1986).

Vickers, Sally. *Miss Garnet's Angel*. Carol & Graf (2000).

Von Schiller, Frederick. *The Man Who Sees Ghosts*. Pushkin Press (2003).

Williams, Jim. *Scherzo*. Scribner (1999).

Winterson, Jeanette. *The Passion*. Grove Press (1987).

Index